'The critical role of non-state actors from communities to social movements, and trade unions and cities, in accelerating green transitions, has been neglected. This volume makes an important contribution by showcasing examples across actors, sectors, regions and issues. It shows how much innovation there is from below driving transitions that are both socially just and environmentally sustainable.'

Peter Newell, Professor of International Relations, University of Sussex, UK

'The corporations have failed us. The politicians have failed us. Who is going to create the sustainable society? This book introduces a great variety of important and realistic sustainability initiatives at the grassroots that go beyond the failing technocratic solutions of the market and the state.'

Pasi Heikkurinen, Lecturer in Management at the University of Helsinki, Finland

'The need to respect planetary boundaries is acutely recognised. This conceptually and empirically rich collection of essays analyses non-state sustainability initiatives that, together, have significant potential to help navigate the complex challenges ahead. This book should be read by everyone with a stake in sustaining life on earth – in other words, all of us.'

Clare Saunders, Professor of Environmental Politics, Environment and Sustainability Institute, University of Exeter, UK

The Role of Non-state Actors in the Green Transition

This book argues that there is no way to make progress in building a sustainable future without extensive participation of non-state actors.

The volume explores the contribution of non-state actors to a sustainable transition, starting with citizens and communities of different kinds and ending with cities and city-networks. The authors analyse social, cultural, political and economic drivers and barriers for this transition, from individual behaviour to structural restraints, and investigate the interplay between the two. Through a series of wide-ranging case studies from the UK, Australia, Germany, Italy and Denmark, and a number of comparative case studies, the volume provides an empirically and theoretically robust argument that highlights the need to develop, widen and scale-up collective action and community-based engagement if the transition to sustainability is to be successful.

This book will be of great interest to students and scholars of climate change, sustainability and environmental policy.

Jens Hoff is Professor in Political Science and leader of the Centre for Sustainability and Society at the University of Copenhagen, Denmark

Quentin Gausset is Associate Professor in Anthropology at the University of Copenhagen, Denmark

Simon Lex is Assistant Professor in Anthropology at the University of Copenhagen, Denmark

Routledge Explorations in Environmental Studies

Environmental Performance Auditing in the Public Sector
Enabling Sustainable Development
Awadhesh Prasad

Poetics of the Earth
Natural History and Human History
Augustin Berque

Environmental Humanities and the Uncanny
Ecoculture, Literature and Religion
Rod Giblett

Ethical Responses to Nature's Call
Reticent Imperatives
James Magrini

Environmental Education in Indonesia
Creating Responsible Citizens in the Global South?
Lyn Parker and Kelsie Prabawa-Sear

Ecofeminism and the Indian Novel
Sangita Patil

The Role of Non-state Actors in the Green Transition
Building a Sustainable Future
Edited by Jens Hoff, Quentin Gausset and Simon Lex

The Creative Arts in Governance of Urban Renewal and Development
Rory Shand

www.routledge.com/Routledge-Explorations-in-Environmental-Studies/book-series/REES

The Role of Non-state Actors in the Green Transition

Building a Sustainable Future

Edited by Jens Hoff, Quentin Gausset and Simon Lex

LONDON AND NEW YORK

from Routledge

First published 2020
by Routledge
2 Park Square, Milton Park, Abingdon, Oxon OX14 4RN

and by Routledge
605 Third Avenue, New York, NY 10017

First issued in paperback 2021

Routledge is an imprint of the Taylor & Francis Group, an informa business

British Library Cataloguing-in-Publication Data
A catalogue record for this book is available from the British Library

Library of Congress Cataloging-in-Publication Data
A catalog record has been requested for this book

ISBN 13: 978-0-367-77715-9 (pbk)
ISBN 13: 978-0-367-23559-8 (hbk)

Typeset in Goudy
by Wearset Ltd, Boldon, Tyne and Wear

Contents

Illustrations

Figures

Tables

Contributors

Hillary Angelo is an Assistant Professor of Sociology at the University of California, Santa Cruz, where she studies the relationship between nature and urbanisation from historical, theoretical and ethnographic perspectives. Her work has been published in leading sociology and geography journals, and her book, *How Green Became Good: Urbanized Nature and the Making of Cities and Citizens*, is under contract with the University of Chicago Press.

Anders Blok is an Associate Professor of Sociology at the University of Copenhagen, Denmark. He currently leads a collective research project on civic urban greening communities in Denmark, combining urban and environmental sociology.

Laura Centemeri is CNRS researcher in Environmental Sociology at the Centre d'Etude des Mouvements Sociaux (EHESS-CNRS-INSERM). Among her recent publications *La permaculture ou l'art de réhabiter* (QUAE 2019).

Anette Gravgaard Christensen is a PhD Fellow at the Department of Sociology, University of Copenhagen. She previously worked as an urban planner. In her PhD, she studies the social dynamics of long-term collaborative planning between municipalities and civil society actors on developing public urban greenspace.

Thomas Hylland Eriksen is a Professor of Social Anthropology at the University of Oslo. He is the author of many books including *Overheating: An Anthropology of Accelerated Change* and *Boomtown: Runaway Globalisation on the Queensland Coast*.

Quentin Gausset is an Associate Professor at the Department of Anthropology, University of Copenhagen. He has 20 years' experience with interdisciplinary research on natural resource management and sustainability. He is currently leading the COMPASS research project (Collective Movements and Pathways to Sustainable Societies). He has co-edited, with Jens Hoff in 2016, the book: *Community Governance and Citizen-Driven Initiatives in Climate Change Mitigation* (Routledge).

Anette Høite Hansen is a PhD Fellow at the Department of Anthropology, University of Copenhagen. She is part of the research project COMPASS that work with Danish collective environmental movements.

Laura Henn is a Research Assistant and PhD candidate at the department of Personality and Social Psychology at the Otto-von-Guericke University Magdeburg, Magdeburg, Germany. Her research interests lie in the field of Environmental Psychology and include the interplay of context factors and individuals' environmental attitude for sustainable behaviour and lifestyles, spillover effects, sufficiency and the promotion of individual sustainability.

Jens Hoff is a Professor in Political Science and leader of Centre for Sustainability and Society at the University of Copenhagen. He has authored, edited and co-edited numerous books, book chapters and journal articles among these *Democracy and Citizenship in Scandinavia* (Palgrave 2001) and *Community Governance and Citizen-Driven Initiatives in Climate Change Mitigation* (Routledge 2016 (ed.)). He is currently engaged in a research project on Collective Movements and Pathways to a Sustainable Society (COMPASS) and a project on the Green GDP.

Mine Islar is a Senior Lecturer at Lund University Center of Sustainability Studies (LUCSUS). Her work analyses the impact of contemporary movements in Europe in creating collective citizen initiatives. She also acts as an expert for the UN Intergovernmental Platform on Biodiversity and Ecosystem Services (IPBES).

Florian G. Kaiser is a Professor of Personality and Social Psychology at the Otto-von-Guericke University Magdeburg, Magdeburg, Germany. His research interests include attitude theory, attitude–behaviour consistency, the Campbell Paradigm, person–situation interaction, evidence-based psychological policy support, large-scale attitude change and behaviour management primarily in environmental protection research. He has published nearly 90 articles in refereed journals and, from 2017 to 2018, he was Co-Chief Editor of the *Journal of Environmental Psychology*.

Jakob Laage-Thomsen is a PhD student at the Department of Organization at the Copenhagen Business School, working on behavioural insights in public policy. Before beginning his PhD, he was part of a research project on civic urban greening communities in Denmark.

Simon Lex is an Assistant Professor at the University of Copenhagen. His research areas are sustainable communities, organisational change and new technologies. He is currently a co-PI in a larger transdisciplinary project on smart technologies and future sustainable cities.

Henrik Hvenegaard Mikkelsen wrote his PhD on mountain communities in Southeast Asia and has recently published the monograph 'Cutting Cosmos: Masculinity and Spectacular Events' at Berghahn.

Clark A. Miller is a theorist of democracy among techno-humans. He is Professor and Director of the Center for Energy and Society at Arizona State University. His most recent books include *The Weight of Light: A Collection of Solar Futures*; *Designing Knowledge*; *The Handbook of Science and Technology Studies* and *Science and Democracy: Knowledge and Power in the Biosciences and Beyond*.

Anne Bach Nielsen is a PhD Fellow at the Department of Political Science, University of Copenhagen. Her PhD investigates how cities respond to increasing risks stemming from climate change with a particular focus on how cities take political decisions through participation in transnational networks.

Stephen Pollard is a PhD student at the University of Melbourne whose research examines local climate governance and net zero emissions. In 2018 he was awarded an Australian Endeavour Research Fellowship to undertake fieldwork in Denmark, hosted by the University of Copenhagen. He has a background in anthropology and has worked in public administration on urban planning, environment and climate policy.

Vivian Price PhD is a Professor at California State University, Dominguez Hills in Interdisciplinary Studies and Labour Studies, and is active in social justice unionism and the struggle against environmental racism. She is a filmmaker whose award-winning work includes *Hammering It Out* (2000), *Transnational Tradeswomen* (2006) and *Harvest of Loneliness* (2010), and has published numerous peer-reviewed articles on gender, labour, visuality and pedagogy. She was a Spring 2018 Fulbright Scholar at the University of Liverpool studying labour and climate change.

David Wachsmuth is the Canada Research Chair in Urban Governance at McGill University, where he is also an Assistant Professor of Urban Planning. He co-directs the Adapting Urban Environments for the Future cluster of the McGill Sustainability Systems Initiative, and is the Early Career Editor of *Territory, Politics, Governance*.

Acknowledgements

First and foremost we wish to thank the Faculty of Social Science, and in particular the Department of Political Science, at the University of Copenhagen, who encouraged and financed the establishment of a Centre for Sustainability and Society, which opened in June 2018. These institutions also financed the opening conference of the Centre, which took place from 6 to 8 June 2018.

A part of the opening conference was a Book Symposium and a PhD seminar, which were financed by the departments of Anthropology, Political Science, Psychology and Sociology. All chapters in this book are based on papers presented at these events. We would therefore also like to thank the authors of the papers for their willingness to transform their work into chapters for this book, and for their patience during the long editing process.

We also wish to thank our research assistants, Bothilde Nielsen and Delara Christensen, who assisted during the conference, and Venil Sælebakke who was indispensable in preparing the manuscript for publication.

Finally, we are grateful for our contacts at Routledge: Editor Annabelle Harris and Editorial Assistant Matthew Shobbrook, for all their encouragement and help in getting this book to the final stage of production. And last, but by no means least, a special thanks to Susan Michael, who did a thorough linguistic revision and copy editing of all chapters in the book.

Jens Hoff
Quentin Gausset
Simon Lex

1 Introduction

Jens Hoff, Quentin Gausset and Simon Lex

Building a sustainable future is one of the greatest and most urgent challenges of contemporary societies. However, how this goal can be realised is still very unclear. We know that past and present approaches to reduce global warming have only had limited success, and that policy makers, public and private organisations, as well as researchers are much in doubt about which paths must be taken to build sustainable societies. This book argues that there is no way to make progress politically, theoretically or on the ground without an extensive participation of non-state actors such as cities, private companies, unions, eco-societies, food cooperatives, urban gardening projects, permaculture initiatives, stop-wasting-food campaigns, and so on. These actors are trying to create concrete pathways to more sustainable futures, and their work must be studied and analysed in order to learn from their successes and failures.

Stressing the role of non-state actors in the transition to low-carbon and more sustainable societies is nothing new. Indeed, at the Rio Earth Summit in 1992 it was suggested that local governments, civil society groups and private organisations would need to be responsible for the lion's share of greenhouse gas mitigation, and the shift towards more sustainable societies; an idea ratified by the passing of the Agenda 21 treaty (Connolly *et al.* 2012). While Local Agenda 21 activities have had a very different fate in different countries – in some being forgotten, in others being institutionalised and integrated in legal frameworks – the focus on non-state actors came to the fore again after the failure of governments to reach a binding agreement on greenhouse gas emissions at the COP 15 meeting in Copenhagen in 2009 (Hoff and Gausset 2016). Following this impasse by the world's political leaders, cities, local governments, civil society and market actors became leaders in the drive to reduce greenhouse gas emissions and create more sustainable societies. Despite the success of actually reaching a binding agreement on maximum global temperature rise in Paris 2015, and the subsequent success of the UN in passing the Sustainable Development Goals (SDGs) in 2016, non-state actors are the indispensable companions for any government wanting to realise the SDGs, and non-state actors still seem to lead the way.

With this book we try to cover this vast research field as best we can. We therefore present a series of case-studies from the UK, Europe (North and

South) and Australia as well as some studies of cities and municipal networks around the world. In terms of the actor focus, this book is very diverse as it focuses both on the individual and the determinants of his/hers (un)sustainable behaviour, on communities of various kinds, analogue as well as virtual, and the personal as well as wider social and political impact. Finally, it also focuses on cities trying to become sustainable or resilient and the importance of techno-social infrastructures. The actor focus structures the book, starting with a focus on 'Individual and collective sustainable norms and behaviour' (Part I), then moving on to 'Grassroots, green communities and social impact' (Part II), and finally dealing with 'Creating sustainable cities and infrastructures' (Part III). Some might argue that this diversity in terms of geography as well as actor focus limits the possibilities of generalizing from the cases and upscaling the 'best case' examples. However, we find such an approach to a discussion of the wider societal importance of the cases included here rather misleading. Thus, many of our cases seem to refute the idea of scales, as they represent (larger or smaller; local–global) networks in which ideas and actors circulate physically and virtually. So in the language of scales, upscaling, downscaling and rhizomatic growth seem to be taking place sequentially or simultaneously. Such reality is much closer to the theoretical approach suggested by, in particular, actor-network theory (see Latour 2005; Jasanoff 2010; Blok 2010), than to, for example, theories on multi-level governance (see for example Bulkeley and Newell 2010). The reader will therefore also find that the question of upscaling is discussed under other headings in this book, such as, for example, 'traces' (Chapter 5), 'diffusion' (Chapter 6) or 'reciprocal exchange' (Chapter 8).

We have decided to write this book on the role on non-state actors in creating more sustainable societies because there are few books that deal at length with this issue in all its diversity, and because we believe that community-based engagement is key to a sustainable future. Furthermore, much of the existing literature treats citizens in a 'Homo economicus' perspective, creating a focus on national climate policies and sustainability initiatives, and on incentives to affect individual behaviour through taxes or subsidies or, more sophisticatedly, through nudging (Sunstein 2014). With this volume, we question this approach and explore the potentials of community-based approaches; we explore and discuss how different kinds of communities create and reproduce new norms and values changing individual behaviour. Our approach therefore questions existing paradigms for creating public policies, and points towards citizen-, community- or municipally driven projects and alliances between these as arenas for learning, and for engaging with policy makers. This gives citizen science and citizen participation a more central role in public sustainability policies than is most often the case, but which is necessary if one wants to see real change on the ground.

Theoretically, we move the current focus on individual choices and calculations to collective choices and actions. In investigating this 'paradigm shift' we draw on relevant theories within anthropology, sociology, psychology, political science, and urban planning. For the theoretically inclined reader, it will be of interest to see how these theories are applied and discussed, and maybe especially how the cases presented challenge some of these theories, and create a need for further theoretical work. Instead of listing all the theories used in this volume, we think it more useful to sketch some of the conceptual discussions, which run across the chapters, and which in some cases have challenged existing theories and lead to the development of new conceptual tools.

There are five important issues, which are discussed across many of the chapters. The conceptual development around these issues represents one of the most important results of this book. These issues are: (1) sustainability as a concept and a practice; (2) do-it-yourself culture/'practivism'; (3) the character of eco-communities; (4) effects, impacts or 'upscaling'; (5) learning from participation.

Sustainability

The most well-known definition of sustainability is probably the one found in the Brundtland Commission's report *Our Common Future* from 1987 (WCED 1987). In this report, sustainable development is defined as 'development that meets the needs of the present without compromising the ability of future generations to meet their own needs'. The reason for the prominence of the definition to this day is no doubt the fact that the concept of sustainable development was adopted by the international community, and it laid a basis for the UN Rio Earth Summit in 1992, where states agreed on the guiding principles and a policy framework for sustainable development, laid down in such documents as the Rio Declaration on Environment and Development, Agenda 21 as well as such conventions as the United Nations Framework Convention on Climate Change (UNFCCC), The United Nations Convention on Biological Diversity (UNCBD) and the United Nations Convention to Combat Desertification (UNCCD). All of these texts were influenced by the Brundtland report, both in terms of how the problems were understood, as well as how they should be tackled (see Stevenson 2018, 123).

Since Rio there has been an ongoing discussion as to what extent the Brundtland Commission's definition of sustainability represent a so-called 'weak' or 'strong' version of sustainability. Weak sustainability is based on the idea that protection of the environment and our natural resources is possible within our current economic and political structures, and that economic growth and sustainability is compatible. Thus, by enhancing efficiencies in production and consumption, through innovation and new technologies and the right incentive structure for companies and individuals, it is possible to create win-win solutions that allow economic growth to continue, and create new market opportunities (Bulkeley *et al.* 2013, 964). Strong sustainability, on the other hand, argues that it is not possible to address the drivers of ecological damage and social inequality without fundamental economic and social reforms (Stevenson 2018, 126). While there is little doubt that most of the world's governments, private companies, and even NGOs have adhered to an interpretation of the

Brundtland Commission, along the lines of weak sustainability, or what Hajer has called 'ecological modernization' (Hajer 1995), and what most decision-makers today would call 'green growth', it is an open question as to whether such interpretation is in fact embodied in the Brundtland definition. Langhelle (2010) for example insists on a more radical interpretation of the Brundtland definition, and points to the fact that the report sees economic growth only as environmentally and socially sustainable if there is a change in the concept of growth to make it more equitable in its impact; i.e. to improve the (re-)distribution of income. He also points to the hierarchy of priorities within the conception of sustainable development in *Our Common Future*, where the first priority is the satisfaction of human needs, in particular the essential needs of the world's poor to which overriding priority should be given, then climate change (and thus the energy issue), loss of biological diversity, pollution (PCB, acid rain, etc.) and food security (Langhelle 2010, 411).

While doubts can be raised about the correct way to interpret the Brundtland Commission's definition of sustainability, there is less doubt about the fact that little progress was made at the nation level on actually creating more sustainability in the first decades after the Rio Summit in 1992. Thus, when 191 governments meet at the World Summit on Sustainable Development in Johannesburg in 2002 to assess progress and discuss new ways of advancing sustainable development, it was quite clear that states had done little over the past decade to implement promised actions. Furthermore, 'unable to report on any significant accomplishments, governments turned their attention to creating "partnerships", which in practice delegated much of the responsibility for sustainable development to the private sector and civil society' (Stevenson 2018, 128). This story was, by and large, repeated at the Rio+20 Summit Meeting in Rio 2012, where the world's nations met again to take stock of the progress made concerning sustainable development at the global level. A report was made prior to the meeting by the Stakeholder Forum (together with the UN) to evaluate how the international community had performed with regard to the goals set at Rio in 1992. In general, the findings were disappointing. Thus, out of 40 topics no progress or even regress was made on most topics; most notably consumption, atmospheric protection and land management. However, some areas had good progress, such as toxic chemicals management, scientific research and inclusion of local authorities in sustainable development governance (Stevenson 2018, 133). The most promising outcome of the summit meeting was that the international community in the outcome document 'The Future We Want' reaffirmed its commitment to sustainable development and began a process towards defining the so-called Sustainable Development Goals (SDGs), which were adopted by the UN in 2015. The 17 goals and 169 targets to be reached by 2030 might signal a new turn in the global attention towards sustainability even though they have been criticised for being 'unactionable' and 'unattainable'.

While many governments, private companies, NGOs and others have embraced the SDGs, at least rhetorically, the SDGs have also – and maybe more interestingly – led to new conceptualisations of sustainability. One such dominant interpretation takes it point of departure in the concept of the Anthropocene; meaning a new geological epoch, where humans are the driving force behind planetary changes – such as depletion of natural resources, pollution of the global commons and global warming. Researchers such as Johan Rockström and colleagues (2009) have identified and quantified a number of planetary boundaries that must not be transgressed if we want to avoid unacceptable environmental change, and they have pointed out that these boundaries have already been transgressed concerning such critical factors as biodiversity, the nitrogen cycle and climate change. They therefore propose a framework of 'planetary boundaries' aimed at defining 'the safe operating space for humanity'. While these planetary boundaries can be said to define an 'ecological ceiling for economic growth' if it is to be sustainable, an economist such as Kate Raworth (2017) has coupled the planetary boundaries with the SDGs. With these goals, the 193 member-states of the UN have committed themselves to eradicate poverty, end hunger, ensure education and decent jobs and adequate health services for all (just to mention some of the 17 goals). These goals can be said to constitute a *social bottom* for sustainability. So, according to Raworth, economic growth can only be seen as sustainable if it takes us somewhere between this ecological ceiling and the social bottom, to what she calls 'the safe and just operating space for humanity' (or into what she calls 'the doughnut').

Some might argue that this newer way of understanding sustainability is just a more precise specification of the Brundtland definition, while others might argue that it is definitively more radical, as Raworth directly encourages us to be 'agnostic about growth' (Raworth 2017, 243ff.) meaning that ecological boundaries and social and environmental justice must be prioritised before and over economic growth. However, what unifies the two approaches is that they both operate at a macro-level, and are therefore relatively unclear about exactly what paths we as societies, communities or individuals need to follow in order to become fully sustainable. This leaves ample room for numerous understandings of sustainability, and for numerous approaches to sustainability in practice. What we see across the cases in this book confirms this point. Thus, even though most of the actors in our cases – certainly cities and bigger and smaller communities –know and claim to work according to the SDGs if asked, the SDGs are relatively remote and hazy goals for most. Instead our actors define the concept performatively: through their practices and communication. For them, sustainability is more a process than an end goal, and sustainability remains an open idea, which is conceptualised as a 'potentiality' (Lex and Mikkelsen, Chapter 8, this volume) or an 'imaginary' (Pollard, Chapter 12, this volume). This also explains how, using the discursive flexibility of the concept, cities can be (discursively) transformed from 'sustainability problems' to 'sustainabilty solutions' (Angelo and Wachsmuth, Chapter 11, this volume).

Do-it-yourself and practivism

One recurrent theme in the chapters of this book is the desire to do something without waiting for national governments or the international community to act. This desire can be found in individuals (see Henn and Kaiser, Chapter 2, this volume), but can also take the form of collective action, which can take place at different scales, ranging from cities, to unions, to groups of citizens who meet in real life or who interact through social medias. What is common to the majority of cases described in this volume is that collective action is not geared towards influencing policy makers upward, but is rather concerned with taking responsibility, taking matters into their own hands, and changing things locally. While classical environmental movements, and the scientific literature that describes them, focus on activism and lobbyism (see for example Jamison *et al.* 1991; Kamieniecki 1993; Wapner 1996; Rootes 1999a, 1999b; Shabecoff 2003; Doyle 2005; Karan and Suganuma 2008), the new type of environmental movements described in the current volume take the form of environmental communities, and focus rather on what can be called 'practivism'.

Practivism is here not understood as the practical methods of doing activism (a definition used on www.practivism.ca, for example), but is rather defined as a new form of activism that is based on changing one's own practices rather than changing those of others. The expansion of practivism derives from the failure of national and international governments to satisfactorily tackle environmental challenges, and thereby also from the failure of classical activism in pushing these government to do more. It also goes hand in hand with a shift in focus from protecting nature and biodiversity to mitigating global warming. While the bad guys destroying biodiversity are conventionally seen as multinational companies exploiting restlessly fragile natural resources, the culprits responsible for global warming are commonly described as each and every one of us, lay consumers whose lifestyles heat up the globe. In other words, while traditional environmental movements are fighting global external enemies, the new environmental communities are fighting to change their own behaviour and consumption patterns internally and locally. In doing so, they also practice a new form of political engagement, which is much more local and inwardly oriented than in the past.

In the current book, practivism is found at a variety of scales, from loose communities built around social media (Lex and Mikkelsen, Chapter 8, this volume), to green eco-communities or cooperatives (Høite Hansen, Chapter 3; Gausset, Chapter 4; Hoff and Islar, Chapter 5; Gravgaard Christensen *et al.*, Chapter 7). Saying that practivism is also found in cities (discussed by Angelo and Wachsmuth, Chapter 11; Pollard, Chapter 12; and Nielsen, Chapter 13) might be pushing the argument a bit far. Cities have always been inward-looking and are constantly trying to change their own practices. But a common ground between the green communities and the cities studied here is the fact that individual actors (whether physical individual persons, or juridical persons/cities) acknowledge the role they play in global warming, and the responsibility that they have in changing things, not just for their own benefit, but also for the benefit of the globe. Seen from this point of view, green communities are on the same practivist wavelength as cities working for a sustainable transition. Likewise, Miller (Chapter 14) argues that the current focus on sustainability pushes the highly technical field of energy management, which is traditionally driven in a top-down manner, towards a more bottom-up approach of socio-technological self-governance.

The new practivism of green communities differs therefore from the old activism of environmental movements exemplified in this volume by the anti-fracking case of Price, (Chapter 10) or the anti-mining movement described by Eriksen (Chapter 9, even though Eriksen argues that the question that people are asking themselves is not just 'Who should I blame', but also 'what can I do' in practice). Of course, the divide between the old activism and the new practivism is not as clear-cut as we present it. In reality, there is a continuum between the two ideal types, and many people are involved in both activism and practivism. Yet, it can be argued that the gravity centre is slowly being displaced from a focus on demonstration and opposition to a focus on own positive action. Centemeri (Chapter 6) shows an example of this transition, as environmentalist movements focus increasingly on practising ecological care, and on repairing environmental degradation rather than on just opposing it.

The apparition and development of practivism brings new research focus and questions regarding the importance of micro-politics, the origins of environmental attitudes and motivations, the interplay between citizens and local governments, the collective aspects of behaviour change, the pragmatism required to live sustainably, and the link between community and commonality. The different papers in this anthology discuss these themes and draft a variety of theoretical arguments allowing us to capture their novelty.

Community

A predominant tale in scholarly debates concerns how the community will serve as a vector for a sustainable environmental transition. As an alternative to rational and individualistic environmental policies, which are fundamentally founded on an understanding of behavioural change as a consequence of individual cost-benefit assessments, non-state actors have, in recent decades, turned to collective actions in order to generate pro-environmental changes. In what could be termed as environmental communities, as suggested by Gausset (Chapter 4), citizens now take matters into their own hands by collectively transforming daily consumption patterns towards a more socially interdependent and sustainable lifestyle. In such collective action-driven change, citizens band together in regular interactions with a common aim of making a positive environmental impact in the local (and global) society. As an example of an environmental community, eco-villages form an alternative physical and socio-political infrastructure that facilitates a self-sufficient sustainable lifestyle. In 'thick' and devoted social groups, people build small villages based on particular social norms and ideals, which are, for example, direct democracy and socioc-racy. Moreover, they are composed of environmentally friendly infrastructures, including houses made by bio-degradable materials, ecologically-based purifying sewage systems and local 'off-the-grid' renewable energy production and con-sumption. These experimental villages thrive on passionate and visionary indi-viduals, and in order to materialise the ideals by practically building and maintaining the alternative way of life, it requires a strong social group that pools together resources such as time, know-how, capital and influence.

While the eco-villages typically reside in rural areas, large-scale environ-mental changes are underway in cities, as political strategies and development goals on how to generate a sustainable society, in recent years, have entered urban planning and policy governance. Environmental or 'green' communities, such as for example food collectives, small urban gardens and biodiversity efforts, arise in cities around the world. Inspired by Laurent Thévenot's 'com-monality in the plural', Christensen, Laage-Thomsen and Blok (Chapter 7) explain that such green engagements are co-composed across place attachment, neoliberal governance and a political-theoretical notion of civic 'green' solid-arity and responsibility. For instance, urban gardens, which belong to a certain geographical location in the city, are in active civic engagements, typically on the part of local citizens, developed with an ambition of pushing the socio-material environment towards a 'greener' future. The gardens, it seems, convert broadly construed political visions into communal and practice-based engage-ments, opening them for continuous internal and external civic-political nego-tiations. As Christensen, Laage-Thomsen and Blok suggest, the urban community furthermore involves 'a set of non-place-specific issues, commit-ments, and concerns [that] extend beyond "the human" into issues of how urban-based technologies and natures become sites of socio-political struggle as well as new forms of affect and solidarity'. Thus, the urban communities, in both practical work in the garden and in visionary meetings of a future sustainable city, turn into arenas where global environmental political narratives sit side-by-side with more ordinary, local activities.

Communities are often marked in relation to their *boundaries* (cf. Barth 1969; Cohen 1985): the eco-village and the urban garden are, for example, *altern-atives*, defined in shared practices and in demarked physical and ideological boundaries, which stand in contrast to 'ordinary' conditions in the surrounding society. Different from these communities, people and organisations also associ-ate on digital platforms and infrastructures. On social media sites, citizens with common pro-environmental aims engage in fluid 'rhizomatic' networks that continually emerge and fuse together. The digital structure facilitates an oppor-tunity for individuals, associations and corporations to share information about the cause, and to mobilise people to follow and participate in activities and events. Furthermore, the sharing of know-how, good practices and advice helps, according to the people administrating the sites, to inspire people to reflect on their consumption patterns, 'pushing' them to change their behaviour in more environmentally friendly directions. While community usually refers to an actual group of people interconnected through a shared sense of belonging, a common 'we', marked by particular physical and symbolic boundaries, the logic of this type of community, as suggested by Lex and Hvenegaard Mikkelsen (Chapter 8), does neither require physical attachment or social containment of a shared identity, nor is it defined in relation to other current groups. Rather, it is loosely composed around an urgent desire to bring about particular sustainable futures. While scholars such as Benedict Anderson (1983) and Anthony Cohen (1985) portray the community as synchronically constituted through internally-shared symbols that are reinforced in relation to other, external, contemporary entities, the contingency of the community is in this case diachronic, as it extends in time towards a potential future outcome. By the advent of pro-environmental online groups, Lex and Hvenegaard Mikkelsen present digital communities as: 'more than anything assemblages of potentials that are perpetually on the brink of being realised'.

Impact and scalability

In the post-industrial individualisation, the grand political discourse seems to have been left behind (Jacobs 1999). As suggested by Phil Macnaghten (2003), a dominant and normative environmental narrative has engendered detached ideas that make little sense to people in their everyday endeavours. This, according to Macnaghten, has made it politically and practically difficult to reach deep and extensive pro-environmental impacts in contemporary societies. However, we now, more than ever, see how non-state actors engage in local environmental actions and politics, in digital and physical spaces, as expressions of individual concerns and lifestyles (Bennett and Segerberg 2012). In this, environmental communities and networks are attached to and formed by particular shared places, practices and ideas, yet, the material and immaterial cornerstones of the community, as argued by Cohen (1995), are multivocal, meaning that they are open for an individual's diverse interpretations and ways of life. For this reason, the community, imagined or practised, integrates collective ideas with individual preferences, making it a strong instrument for advancing sustainable transitions on local and regional scales.

Gausset acutely summarises the influence of environmental communities in four categories: '(1) through designing physical, social, and political infrastructures that influence everyday-life choices; (2) through collective dynamics characteristic of collective identities in local communities; (3) through producing and transferring knowledge and experience; and (4) through political engagement and lobbying.' As a consequence of joining environmental communities, he suggests, people tend to become more willing to explore new and sustainable lifestyles. As well as being experimental arenas of growing environmental collective actions, these communities also have an explicit aim for transferring practical knowledge to other people and organisations in society. As examples, collective experiences and ideas are shared with people from all over the world in inspirational tours and visits in the eco-villages and in teaching courses in open universities ('højskoler'). This, we believe, opens possible pro-environmental impacts in new similar collectives formed by people in other regions. Along these lines, Hoff and Islar (Chapter 5), explain how a food cooperative in Copenhagen has directly stimulated larger supermarket chains to develop new successful sustainable commercial products. Furthermore, the cooperative has collaborated with local schools, making deliveries of vegetables to the school canteen, and cooking meals and sharing experiences with teachers, pupils and parents.

Furthermore, it is important to point out that environmental communities also are used by influential actors to promote sustainable ideas and goals in political arenas. The founder of the Stop Wasting Food Movement in Denmark, for instance, has attained access to political leaders through her community of tens of thousands of 'ordinary' and renowned citizens. Seemingly, she uses her loosely tied digital network to gain influence in the political sphere, making it an indirect vehicle for forging constructive impacts on policies and legislation. By becoming a part of a strong *brand* in interactions with the community activists, the political actors gain access to a positive public narrative. The interactions between the grassroots and the political elite in this case unfold in a continuous reciprocal accumulation of symbolic capital, where the involved partners gain value by joining together in the fight against food waste. Likewise, current and former members of the food cooperative are involved in mainly left-wing political parties in which they have directly developed political strategies inspired by the cooperative´s principles of transparency, inclusiveness, and responsible local food production and consumption.

These cultural, social and material impacts engendered by environmental communities in local and global societies could be termed as perpetual *traces* towards sustainable futures.

Moreover, in relation to the changing of people´s lifestyle patterns, Lex and Hvenegaard Mikkelsen (Chapter 8), inspired by Bateson (2000), argue that potential future impacts have the ability to unlock the present by offering hope and ambition to overcome pressing and seemingly unmanageable environmental concerns and challenges. Thus, while environmental actions, as mentioned, in imagined and practising communities, transgress the contention between *the individual* and *the collective*, and leave impactful collective traces in continuing proceedings from concrete groups into society at large, they also extend in time, embroiling future potentials into concrete transformative actions in the here and now.

Knowledge and learning from participation

No one knows exactly how a sustainable transition can be achieved, but what is certain is that our sustainable future will depend on a mix of new technologies, new forms of governance, and new lifestyles and consumption patterns, and that all solutions must be in place within the next 12 years if we want to have a fair chance to keep global warming below 2° Celsius, compared with pre-industrial times. Will technologies be developed fast enough? How can we reinvent our democracies, collective decision-making mechanisms and labour relations? Will people accept reducing their consumption and transport, and change their eating habits? Can we do this by, or while, securing a better redistribution of wealth? Again, no one knows for sure.

What is fascinating, however, and a source of great hope, is that an increasing number of communities are experimenting, trying, failing, trying again, and designing in this process new technologies, new forms of governance, and new lifestyles. They are, indeed, reinventing the world, even though they start doing this at the scale of a green cooperative, an eco-village or a transition city. In other words, these communities are engaging in an exciting knowledge production and learning process in which creativity and experience are keys. They are also creating networks and alliances to exchange and spread their knowledge and experiences, which generates ripple effects. All this contributes to building a momentum susceptible to change the world and 'business as usual', if it can acquire a critical mass that is large enough to reach a tipping point.

Knowledge and learning are here deeply ingrained in practices. Eco-communities and transition cities constantly meet challenges, and constantly devise new pragmatic solutions to overcome them. The knowledge produced in this process is, in the terminology of James Scott (1998), a mix of *techne* and of *metis*. On the technological side, these communities are the first to implement the newer and most sustainable technologies relating to renewable energies, for example. By being pioneers in the implementation of such technologies, they produce invaluable experience and feedback, which gives them an important role to play in technological development. When it comes to metis, important knowledge is developed and acquired through the everyday practice and experience of running a sustainable community. Such practical knowledge is transferred through learning by doing rather than as dogmatic knowledge spread through blueprints.

Diversity is a striking feature of community-based solutions. Each community develops or appropriates knowledge and technologies that are adapted to local context, creating in this process a mosaic of experiences, relating to renewable energy (see Gausset, Chapter 4; Miller, Chapter 14), new forms of local or micro-governments (Angelo and Wachsmuth, Chapter 11; Christensen *et al.*, Chapter 7; Hoff and Islar, Chapter 5; Nielsen, Chapter 13; Pollard, Chapter 12), behaviour change and new lifestyles (Gausset, Chapter 4; Hansen, Chapter 3; Henn and Kaiser, Chapter 2; Hoff and Islar, Chapter 5; and Lex and Hvengaard Mikkelsen, Chapter 8), or else relating to classical anti-exploitative movements (Centemeri, Chapter 6; Eriksen, Chapter 9: Price, Chapter 10). Some chapters have an explicit and specific focus on how knowledge and experiences are produced and shared, and demonstrate the innovation and creativity characterising the movement. These new experiences are never 100 per cent replicable as such, but can always inspire similar or newer pilot projects elsewhere, once they are adapted to local context. They thereby make an important contribution to sustainable futures.

References

Anderson, B. (1983) *Imagined Communities: Reflections on the Origin and Spread of Nationalism*. London: Verso.

Barth, F. (1969) *Ethnic Groups and Boundaries: The Social Organization of Culture Difference*. London: George Allen & Unwin.

Bateson, G. (2000) *Steps to an Ecology of Mind*. Chicago: University of Chicago Press.

Bennett, L. and Segerberg, A. (2012) The logic of connective action. *Information, Communication & Society*, 15(5), 739–768.

Blok, A. (2010) Typologies of climate change: actor-network theory, relational-scalar analytics, and carbon-market overflows. *Environment and Planning D: Society and Space*, 28, 896–912.

Bulkeley, H. and Newell, P. (2010) *Governing Climate Change*. London and New York: Routledge.

Bulkeley, H. *et al.* (2013) Governing sustainability: Rio+20 and the road beyond. *Environment and Planning C: Government and Policy*, 31, 958–970.

Cohen, A. (1985) *The Symbolic Construction of Community*. London: Routledge.

Connelly, J., Smith, G., Benson, D., and Saunders, C. (2012) *Politics and the Environment: From Theory to Practice*. New York and London: Routledge.

Doyle, T. (2005). *Environmental Movements in Minority and Majority Worlds*. New Brunswick: Rutgers University Press.

Hajer, M.A. (1995) *The Politics of Environmental Discourse. Ecological Modernization and the Policy Process*. Oxford and New York: Oxford University Press.

Hoff, J. and Gausset, Q. (eds) (2016) *Community Governance and Citizen-Driven Initiatives in Climate Change Mitigation*. New York and London: Routledge/Earthscan.

Jacobs, M. (1999) *Environmental Modernisation: The New Labour Agenda*. London: The Fabian Society.

Jamison, A., Eyerman R. and Cramer, J. (1991). *The Making of the New Environmental Consciousness: A Comparative Study of Environmental Movements in Sweden, Denmark and the Netherlands*. Edinburgh: Edinburgh University Press.

Jasanoff, S. (2010) A new climate for society. *Theory, Culture & Society*, 27(2–3), 233–253.

Kamieniecki, S. (ed.) (1993). *Environmental Politics in the International Arena. Movements, Parties, Organizations, and Policy*. Albany: State University of New York Press.

Karan, P.P. and Suganuma, U. (2008). *Local Environmental Movements*. Lexington, KY: University Press of Kentucky.

Macnaghten, P. (2003) Embodying the environment in everyday life practices. *Sociological Review*, 51(1), 62–84.

Langhelle, O. (2010) Why ecological modernisation and sustainable development should not be conflated. In A.P.J. Mol, D.A. Sonnenfeld and G. Spaargaren (eds), *The Ecological Modernisation Reader. Environmental Reform in Theory and Practice*. London and New York: Routledge, 391–417.

Latour, B. (2005) *Reassembling the Social. An Introduction to Actor-Network-Theory*. Oxford, UK: Oxford University Press.

Raworth, K. (2017) *Doughnut Economics. Seven Ways to Think Like a 21st-Century Economist*. London: Random House Business Books.

Rockström, J. *et al.* (2009) A safe operating space for humanity. *Nature*, 461(24 September 2009), 472–475.

Rootes, C. (ed.) (1999a) *Environmental Movements – Local, National and Global*. London: Frank Cass.

Rootes, C. (1999b) Environmental movements: from the local to the global. *Environmental Politics*, 8(1), 1–12.

Scott, J.C. (1998). *Seeing Like a State: How Certain Schemes to Improve the Human Condition Have Failed*. New Haven: Yale University Press.

Shabecoff P. (2003) *A Fierce Green Fire: The American Environmental Movement*. Washington, DC: Island Press.

Sunstein, C.R. (2014) *Why Nudge? The Politics of Libertarian Paternalism*. New Haven and London: Yale University Press.

Stevenson, H. (2018) *Global Environmental Politics. Problems, Policy and Practice*. Cambridge, UK: Cambridge University Press.

Wapner, P. (1996) *Environmental Activism and World Civic Politics*. Albany: State University of New York Press.

WCED (World Commission on Environment and Development) (1987) *Our Common Future*. Oxford: Oxford University Press.

Part I

Individual and collective sustainable norms and behaviour

2 Sustainable societies

Committed people in supportive conditions

Laura Henn and Florian G. Kaiser

Varying levels of sustainability across societies

Since Western societies consume resources at a rate that exceeds by far the capacity of our planet to replenish them (e.g. McNeill 2000; Steffen *et al.* 2011; Hoekstra and Wiedmann 2014), sustainability[1] requires a massive reduction in energy and resource consumption. There is consensus among nations that steps need to be taken to decrease resource and energy consumption and by doing so to also slow global warming (see United Nations 2015). However, there is a robust debate among climate change scholars and experts about the most efficient methods we should use to reduce excessive energy and resource consumption.

One part of the debate involves the role that individual behaviour plays in achieving sustainable energy and resource consumption levels. Whereas some people have questioned the relevance of individual behaviour altogether (e.g. Grunwald 2010), others have argued that only a small fraction of behaviours are impactful enough to make a difference (e.g. Stern 2000; Bilharz 2008). By contrast, we argue that all human behaviour matters to sustainability, and it is the entire array of behaviours that together contribute to the problematic scale of consumption. Sustainability requires the *sustainable performance* of individuals: various sustainable behaviours have to be implemented continuously and with a certain rigour. Over time, this overall pattern of behaviour will result in lower energy and resource consumption and hence, in lower levels of greenhouse gas emissions.

Even within the European Union, countries differ dramatically in their sustainable performance. This can be seen in a comparison of the different per capita greenhouse gas emissions: whereas Luxembourgians emit 20 tonnes of greenhouse gases per capita per year, Estonians emit 14 tonnes, and Croatians and Swedes only about six tonnes (Eurostat 2014). Of course, per capita emissions are dependent on the geographical and political features of a sociocultural context and the climate and socioeconomic conditions of a given country. Contextual factors jointly affect people's resource and energy consumption, for example, for heating and transportation. However, as the similar emission levels of Sweden and Croatia indicate, climatic and socioeconomic conditions are not deterministic constraints. As a northern country with high energy demands and a high average income, Sweden could have been expected to have much higher emission levels than a warmer, less prosperous country such as Croatia.

Accordingly, Sweden must have structural, man-made conditions that facilitate the comparatively sustainable performance of this country despite its wealthy, northern European characteristics. Thus, countries that have renewable energy sources and also invest in using them will improve their citizens' emission balance and facilitate more sustainable ways of living.[2] For example, sustainable traffic conditions depend to a large extent on public policy-making and are thus also subject to man-made decisions. Improving and extending roads and parking facilities will encourage more car traffic, whereas giving priority to bike infrastructures and expanding public transportation services will encourage more sustainable mobility. Every sustainable behaviour is largely dependent on contextual factors that define its 'costs', that is, its relative difficulty (see also, for example, Thøgersen 2005, 2014; Steg and Vlek 2009; Kaiser *et al.* 2010; Byrka *et al.* 2017; Arnold and Kaiser 2018). Therefore, *context* is an important, albeit not the only, lever that can be applied to promote sustainability.

Moreover, there is quite a bit of variation in the extent to which individuals exhibit sustainable performances. Inhabitants of the same country can exhibit very diverse per capita greenhouse gas emission levels (see, for example, Ivanova *et al.* 2017; German Federal Environmental Protection Agency [UBA] 2018), which implies that there are always ways for individuals to influence their own emission levels. People's per capita emissions are a function of their specific behavioural choices. Taking an aeroplane from Europe to New York and back, for example, can double an individual's annual emissions (German Federal Environmental Protection Agency [UBA] 2018). Maintaining a vegan diet produces only one-half of the diet-related greenhouse gas emissions that a meat-based diet produces (Scarborough *et al.* 2014).

However, a single sustainable behaviour does not make a discernible difference in people's overall sustainable performance. Lower energy and resource consumption requires an overall pattern of sustainable behaviour by, for example, considering low-emission products in *any* purchase, reducing mobility *in general* (and avoiding emission-intensive transportation by car or aeroplane in particular), reducing one's living space, installing weatherisation, and reducing waste generation. In short, people's sustainable performance is the result of the specific ways in which people lead their lives. Hence, as we explain in more detail below, a person's esteem for sustainability and his or her appreciation for environmental protection, that is, their *environmental attitude*, are crucial for sustainable performance.

In the following section, we elaborate on the interplay of individuals' motivation to support sustainability (i.e. their environmental attitude) and the conditions that facilitate or aggravate sustainable behaviour. We also discuss the idea that behaviour can be changed by altering either the conditions surrounding a behaviour or people's environmental attitude. In the subsequent section, we turn to the broader view of sustainable performance and argue that behaviour change alone is insufficient: only people who are motivated to act sustainably will eventually adopt a sustainable lifestyle.

Sustainable behaviour: a function of environmental attitude and behavioural costs

If Emma chooses to bike to work instead of driving by car, there may be many different reasons behind her choice. She might enjoy physical activity for the fun of it, she might wish to avoid parking costs, or she might see biking as the fastest way to get to work. However, if Emma bikes to work even on a rainy day and if she recycles waste, buys organic produce, saves energy, and even donates to environmental organisations, these behaviours have one thing in common: they are relatively favourable for the environment. There is hardly another plausible explanation for why Emma would perform all these behaviours than her high esteem for environmental protection and sustainability (see, for example, Kaiser *et al.* 2010). Her pattern of behaviour can thus be considered an expression of the favourable attitude that Emma holds toward sustainability.

The importance that an individual attaches to sustainability – also called a person's environmental attitude – is reflected in the extent to which that person engages in environmentally friendly behaviours even in the face of behavioural costs (Kaiser *et al.* 2010). Whereas some behaviours require comparably little effort (such as turning off the lights when leaving a room, recycling waste, and not littering), other behaviours are quite demanding (purchasing an electric car), inconvenient (refraining from using a car), time-consuming (traveling long distances by train instead of aeroplane), or even socially challenging (confronting someone who is acting in an environmentally harmful manner).

The higher a person's level of environmental attitude, the more likely he or she will engage in any specific sustainable behaviour. For example, people with a higher environmental attitude are more likely to subscribe to a green electricity tariff (Arnold *et al.* 2018). They are also more likely to be vegetarians (Kaiser and Byrka 2015), to bike to work (Kibbe 2017; Taube *et al.* 2018), to accept the restrictions that come with living near a nature preserve (Byrka *et al.* 2017), to participate in environment-related surveys (Arnold and Kaiser 2018), and to prefer organic products (Taube and Vetter 2019).

The higher the costs of a behaviour, the more likely it is that only people with higher levels of environmental attitude will engage in it. For example, Arnold and Kaiser (2018) asked participants in their study not only a small favour (i.e. to sign a petition) but also a subsequent bigger favour (to fill out more questionnaires). People who agreed to both favours had a significantly higher level of environmental attitude than people who agreed to only the smaller favour. This means that asking people to engage in successively costlier behaviour leads to a systematic dropout of the less motivated. In another study, Taube and Vetter (2019; see also Arnold 2017) manipulated the behavioural costs of choosing organic products in a virtual online grocery store using what is

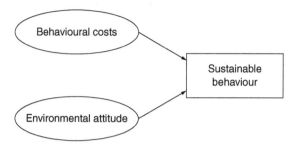

Figure 2.1 Sustainable behaviour is determined by behavioural costs and environmental attitude.

commonly known as 'nudging' (Thaler and Sunstein 2008). They varied the
number of products for which the default option was set to conventional (versus
organic). Participants could either accept the default or click to see an extra
window and choose from a large selection of conventional and organic variations of that product. It turned out that people with higher levels of environmental attitude expended more effort avoiding the conventional default options
and choosing organic products instead. The lower a person's level of environmental attitude, the more likely he or she was to simply accept the conventional
default options instead of putting the extra effort into choosing organic
alternatives.

Whether a person will engage in a specific sustainable behaviour (e.g. subscribe to a green electricity tariff) thus depends on two factors: the objective
costs (which translate into the relative difficulty) of a behaviour and the person's esteem for the environment or sustainability (i.e. the person's environmental attitude; see Figure 2.1).

The two factors independently control behaviour in a compensatory manner:
that is, the higher the costs of a behaviour, the higher the attitude level that is
necessary to overcome these costs. As a consequence, as we explain in detail
next, we can utilise behavioural costs to assess people's environmental attitude
levels.

Identifying attitude

Attitudes become apparent in the face of behavioural costs. Sustainable behaviours that are easy to implement, such as separating waste or turning off lights in
unused rooms, are performed by many people, even people who have only a low
level of environmental attitude. Costlier behaviours, such as refraining from car
use or donating to environmental organisations, require more determination:
typically, only people with a correspondingly high level of environmental attitude engage in such behaviours. Accordingly, by exploring a set of sustainable
behaviours of varying difficulties, we can measure individuals' environmental
attitude. The more, and the more difficult, the sustainable behaviours a person

enacts, the higher is his or her environmental attitude. And vice versa: the more inclined a person is to live sustainably, the more and the more difficult are the sustainable behaviours this person will be willing to engage in. Thus, the behaviours in which environmental attitude is expressed can be ordered transitively (by their behavioural costs) in a pattern that pertains to all people in a given sociocultural context (see Kaiser and Wilson 2004, in press; Kaiser *et al.* 2010, 2018).

Figure 2.2 illustrates the compensatory effect that behavioural costs and environmental attitude have on the probability of engaging in certain sustainable behaviours. The left-hand side depicts the presumed normally distributed levels of environmental attitude in a hypothetical population. The right-hand side lists some examples of sustainable behaviour that are indicators of environmental attitude. The behaviours are ordered by their costs, ranging from minor (the lower end of the continuum) to major (the upper end of the continuum). The cost of a behaviour is derived from the relative number of people who engage in the respective activity: behaviours that are performed by most people can be regarded as low in cost (e.g. recycling glass), and behaviours that are only rarely performed can be regarded as costly (e.g. pointing out unecological behaviours in others).

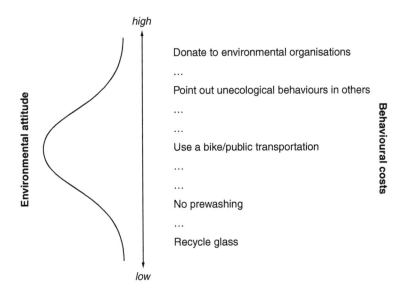

Figure 2.2 The environmental attitude level can compensate for the costs of sustainable behaviour with a corresponding or lower cost level. A person is likely to show a behaviour if his or her attitude level (i.e. the position in the distribution on the left-hand side of the graph) is higher than the costs of the behaviour (i.e. the position along the axis on the right-hand side of the graph). The two factors can be measured in the same metric so they are directly comparable.

Source: authors.

The mathematical relation between behavioural costs and attitudes can be depicted with the Rasch model as:

$$\ln\left(\frac{p_{ki}}{1 - p_{ki}}\right) = \theta_k - \delta_i \tag{1}$$

According to the Rasch model, a person's probability of performing a behaviour – expressed as the natural logarithm of the ratio of the probability (p_{ki}) that person k will engage in a specific behaviour i relative to the counter-probability that person k will not engage in behaviour i ($1 - p_{ki}$; see equation (1)) – is a function of the arithmetic difference between k's attitude level (θ_k) and the composite of the costs (δ_i) involved in engaging in the specific behaviour i.

The two parameters – one for environmental attitude and one for the costs of a behaviour – are expressed in the same metric (i.e. logits) and are therefore directly comparable. With reference to Figure 2.2, this means that individuals whose attitude level (the left-hand side) is higher than the costs of a specific behaviour (the right-hand side, e.g. using a bike/public transportation) are likely to perform that specific behaviour. Individuals with attitude levels that are lower than the costs of a behaviour are not expected to engage in the specific behaviour. When a person's attitude level matches the costs of a behaviour, the person's probability of overcoming the behavioural costs is 50 per cent.

Empirical tests have repeatedly corroborated this model, known in attitude research as the Campbell Paradigm (see, for example, Kaiser et al. 2010, 2018; Kaiser and Wilson, in press). A person's environmental attitude can be measured by recording which sustainable behaviours a person claims to implement, for example, by a self-reporting questionnaire. Hence, this attitude toward environmental protection (and thus toward sustainability) is derived from people's patterns of behaviour, that is, their lifestyles.

According to the logic of the Campbell Paradigm, attempts to increase the probability of people engaging in a specific pro-environmental behaviour (such as biking to work) can target one of two factors: they can aim to either (a) decrease the costs of a behaviour – for example, by removing structural barriers (e.g. installing separate bike lanes) or by reducing the price (e.g. subsidising bike purchases and maintenance), or (b) increase people's levels of environmental attitude – that is, making people more inclined to support sustainability. In the following section, we discuss possible strategies and downsides for both options: decreasing costs and increasing attitude to promote sustainable behaviour.

Promoting behaviour by reducing costs

A straightforward way to increase sustainable behaviours is to reduce the behavioural costs they impose. The implication of such a reduction for a particular behaviour is that it will move down the behavioural cost scale, which is represented on the right-hand side in Figure 2.2. That is, when less dedication to sustainability is required to match the costs, more people will act accordingly. In other words, if we simply make it easier to use public transportation (for example, by waiving any charge or by providing superior service), more people will use it, even those who do not have a very pronounced environmental attitude.

Because behavioural costs are behaviour-specific, each behaviour typically needs to be addressed individually. Some behaviours can be made more accessible (such as wide availability of organic products beyond specialty markets), whereas others can be incentivised (for example, subsidies for installing photovoltaic systems) or socially sanctioned by stressing the behavioural norm (for example, for reusing towels in hotels instead of having them changed daily). Still others can be made less demanding, for example by installing kerbside recycling instead of drop-off systems or by setting double-sided printing as a printer's default.

Reviews and meta-analyses about behaviour change measures in environmental psychology have revealed that such effective measures are characteristically behaviour-specific (e.g. Abrahamse *et al.* 2005; Osbaldiston and Schott 2012; Lokhorst *et al.* 2013; Thomas and Sharp 2013; Schultz 2014; Maki *et al.* 2016). So far, behaviour change measures have been successfully applied to the use of public transportation (e.g. Bamberg 2006; Abou-Zeid *et al.* 2012), waste recycling (e.g. Wang and Katzev 1990; Best and Kneip 2011), reducing paper waste (e.g. Hamann *et al.* 2015), reusing towels in a hotel (e.g. Baca-Motes *et al.* 2012), or reducing meat consumption (e.g. Klöckner and Ofstad 2017). Not surprisingly, behaviour change in environmental psychology typically targets specific behaviours by making these target behaviours either less demanding (by removing barriers to make them easier, e.g. by providing reminders or specific knowledge) or more attractive (for example, by offering financial or social incentives; e.g. Schultz 2014).

For example, recycling could be significantly improved by replacing a drop-off system – where people bring their recyclables to a central collection point – with a kerbside collection system where recyclables are collected from in front of people's homes (see Best and Kneip 2011). Apparently, the effort required to recycle decreased considerably when cities implemented kerbside collection systems instead of requiring people to drop recyclables at a collection point.

On the other hand, extra benefits make the target behaviour more attractive because they compensate for some of the behavioural costs. Predictably, Abrahamse *et al.* (2005) found that rewards are successful strategies for encouraging energy savings. When Germany introduced attractive feed-in tariffs for electricity produced from private photovoltaic installations, it was not surprising that sales of small-scale photovoltaic systems soared (Hoppmann *et al.* 2014). The financial costs of purchasing a private photovoltaic system were compensated by the expected revenues.

Reducing behavioural costs is also the core idea behind nudging (Thaler and Sunstein 2008; see Kaiser *et al.* 2014a): The desirable behaviour is systematically facilitated (e.g. by making it the default or by increasing its salience in other ways). In an experiment on the willingness to participate in a smart grid, for example, the default was manipulated by either offering people to opt in (i.e. the default was 'no smart grid', but people could actively choose to participate in a smart grid) or to opt out (i.e. the default was participating in the smart grid so everyone participated unless they actively refused; see Toft *et al.* 2014). The costs of participating in a smart grid are higher if an active choice is required (opt-in) than if no action is required (opt-out). Accordingly, participation rates for smart-grid participation turned out to be higher for the opt-out compared with the opt-in condition (see Toft *et al.* 2014).

Because cost-reduction measures are behaviour-specific, it becomes essential to find the one that will be most effective for changing a behaviour (see, for example, Steg and Vlek 2009; Osbaldiston and Schott 2012; Schultz 2014). In addition, because such measures typically control only a few behaviours, cost-reduction carries with it several disadvantages and pitfalls when it comes to promoting sustainable lifestyles. First, it appears close to impossible to use cost-reduction strategies to address the infinite number of sustainability-relevant behaviours in people's everyday lives. Second, behaviour-specific approaches bear the risk of undesirable side-effects such as rebound effects or backfiring because people might use their newly freed assets, i.e. time, money, or energy, to try to achieve other goals (see Otto *et al.* 2014). For example, backfiring is apparent when the money saved by utilising smart-meter-based feedback is used to purchase new devices that subsequently consume additional electricity.

Otto *et al.* (2014) explain the unremittingly high energy consumption per capita by illustrating that when sustainability is not a personal goal behind people's behaviour (i.e. when people do not aim to reduce their overall energy consumption and waive additional benefits), energy efficiency measures will simply enable them to reach more yet-unmet personal ends. Such undesired examples of backfiring have been found, among other things, in owners of electric cars who use their cars more often for daily errands than conventional car owners do (Klöckner *et al.* 2013). Similarly, subsidising energy-efficient upgrades of household appliances has led to substantial rebound effects in energy use rather than to the technologically expected energy savings (Yu *et al.* 2013).

To summarise: changing the costs of a behaviour can promote that particular behaviour but does not simultaneously promote others. Nevertheless, many researchers have looked for 'spillover behaviour effects'. Although desirable, evidence of such spillover effects from adopting one sustainable behaviour to subsequently changing other such behaviours is weak, and we believe it is debatable (see, for example, Truelove *et al.* 2014). We argue that spillover will occur only when an intervention effectively improves people's personal commitment (see Henn *et al.* 2019a). Thus, sustainable lifestyles require a change in people's environmental attitude.

Promoting behaviour by increasing people's environmental attitude

If people's environmental attitude could be increased, that is, if the left-hand side of Figure 2.2 were to be shifted 'upwards', a desirable effect would result: all behaviours would become more likely. People will surmount more behavioural costs when their attitude levels are higher, irrespective of the specific behaviour. Unfortunately, attitudes tend to be quite stable over time, and short-term changes therefore seem unlikely (see, for example, Kaiser *et al.* 2014b). However, research has shown that attitudes *do* change – albeit slowly – and that they are specifically susceptible to learning (see Otto and Kaiser 2014). It is thus worthwhile to consider potential ways to alter people's environmental attitudes.

Environmental attitude is correlated with knowledge about the environment and knowledge about effective action for alleviating negative environmental consequences (Roczen *et al.* 2014; Díaz-Siefer *et al.* 2015). Thus, educational measures aimed at imparting knowledge about the environmental system and the influences of behaviour on environmental problems are a promising approach for fostering the formation of environmental attitude in children and adolescents.

Experiences with nature and an appreciation of nature for personal restoration and pleasure are also linked with a higher propensity to protect the environment (see, for example, Mayer and Frantz 2004; Nisbet *et al.* 2009; Brügger *et al.* 2011). People who experience the natural environment as something valuable and worth protecting are more likely to behave pro-environmentally (e.g. Clayton 2003; Mayer and Frantz 2004; Hinds and Sparks 2008; Kaiser *et al.* 2013). Thus, an appreciation of nature potentially helps forge a path to environmental protection guided more by personal benefit and less by selflessness (Kaiser *et al.* 2013). Encouraging children and adolescents to experience nature also potentially promotes a more pronounced environmental attitude and subsequently a more sustainable lifestyle when they grow up.

However, childhood is not the only opportunity for developing esteem for the environment and an appreciation for sustainability. Adults can learn as well and, as a consequence, attitudes can change. Otto and Kaiser (2014) found a small systematic increase in the average level of environmental attitude from 2001 to 2010 in Saxony-Anhalt, one of Germany's states. Because all people, regardless of their age, tended to be more pro-environmental in 2010 compared with 2001, the authors concluded that a learning effect had occurred on the basis of events and experiences. The authors presumed that the change in environmental attitudes was likely fuelled by (probably even indirect) experiences with, for example, hurricanes, floods, and heat waves along with the media's presentation of topics related to climate change and the impact of human behaviour on the natural environment (Otto and Kaiser 2014). However, spontaneous attitude changes occur slowly, and they tend to be modest in size. For the ultimate goal of building sustainable societies, measures that can actively promote substantial changes in people's attitudes are needed.

As we demonstrate in the next section, people's environmental attitude not only compensates for the costs of behaviour and makes sustainable activities more likely, it is also necessary for the ultimate ambition of creating sustainable lifestyles. We argue that environmental attitude is the necessary prerequisite for any behavioural intervention to translate into lower resource consumption levels and is thus crucial for the overall sustainable performance of people.

Environmental attitude: the necessary condition for sustainable performance

A sustainable lifestyle involves an overall sustainable performance made up of a multitude of behaviours that result in sustainable levels of energy and resource consumption. In other words, behaviour changes are successful when they substantially reduce people's levels of consumption. However, as mentioned before, behaviour change interventions are usually targeted toward specific behaviours, which Schultz (2014) calls 'end-state' behaviours, that is, behaviours that directly cause environmental impact, for example, replacing an old with a new energy-efficient refrigerator. However, changing end-state behaviours does not have a large enough impact to contribute to a measurable increase in sustainable performance.

By contrast, behaviour change measures directed at *sets* of behaviours, for example, promoting energy saving or sustainable consumption, are more promising. For example, encouraging people to consult feedback (e.g. from smart meters) on their energy consumption is linked to energy saving behaviours in general. Although consulting smart meters does not save energy directly, it can reduce people's consumption provided people embrace the opportunity to save energy. However, this means that a person must engage in various activities simultaneously and continuously (such as consult the smart meter with some regularity, lower the average room temperature, turn off the lights, and use electric appliances less frequently). Obviously, the success of smart meter feedback depends a great deal on people's commitment (see, for example, Nilsson *et al.* 2014; Webb *et al.* 2014; Gölz 2017). In fact, reviews have shown that all kinds of behavioural change measures are typically more effective for people who are environmentally engaged (e.g. Abrahamse *et al.* 2005; Darby 2006) because behaviour can be implemented in more or less rigorous ways.

Measures that address behaviours that are not end-state behaviours (see Schultz 2014) and thus do not have a direct impact on the environment, but are expected to influence several other behaviours, are promising ways to effectively reduce consumption. For instance, using smart-meter-based feedback does not reduce energy consumption per se; rather, it provides information that *can facilitate* energy saving, but it is an expectedly complex process that converts the information obtained from consulting a smart meter into energy savings. In this process, feedback must be accessed; that is, one might have to register for and log in to a feedback portal where individual feedback is provided. Subsequently, the person must acquire knowledge about actions that are effective for reducing energy consumption. This includes not only knowing that, for example, turning off devices in stand-by mode can help saving energy but also knowing about the relative impact of different behaviours (for example, that reducing the water temperature when doing laundry saves much more energy than turning off devices in stand-by mode). And, of course, one must remember to apply specific energy-saving behaviours in appropriate situations, such as when doing laundry.

We believe that a person's esteem for an object (e.g. environmental protection, low energy consumption levels) or dedication to a goal (e.g. to protect the environment, to save energy) can be seen in all the behaviours a person engages in or avoids. Predictably, the more pronounced a person's attitude, the more attitude-relevant behaviours the person will engage in and the more challenging these behaviours can be (see Kaiser *et al.* 2010). In addition, a more pronounced environmental attitude will also be exhibited in the rigour with which behaviours are implemented. In other words, the efficacy of an intervention that aims to promote energy saving depends on not only *whether* a person pays attention to a smart meter but also *how rigorously* the person engages in the various behaviours that translate feedback into real energy savings.

The rigour with which a person strives to attain a behavioural goal likely depends on this person's esteem for the attitude object (such as protecting the environment or saving energy). Hence, a person's environmental attitude accounts for the rigour with which the person will make use of measures that can help his or her behaviour become sustainable or to protect the environment more effectively. In other words, the sustainable performance (in terms of amount of energy consumed or volumes of greenhouse gases emitted) of a behaviour (such as using smart-meter-based energy feedback, taking public transportation, switching off appliances) is moderated by the person's level of environmental attitude (see Figure 2.3). Thus, only with a pronounced level of environmental attitude can a person be expected to engage in pro-environmental activities rigorously and continuously enough to truly diminish his or her impact on the environment (Otto *et al.* 2014; Urban and Ščasný 2014).

In a field study on the effectiveness of smart-meter feedback for energy saving, we corroborated this attitude-dependent effect of behaviour on people's

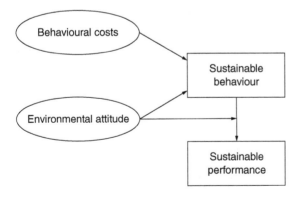

Figure 2.3 Sustainable performance depends on environmental attitude.
Source: authors.

sustainable performance (Henn *et al.* 2019b). We found that customers who voluntarily used smart-meter-based feedback provided by their energy supplier saved energy compared with customers who did not use feedback, but only if they had a rather pronounced level of environmental attitude. In other words, *feedback helps people save energy only when they are already dedicated to protecting the environment.* Or, in more general terms, behaviour will translate into sustainable performance only when people hold a favourable attitude toward sustainability and are hence willing to implement corresponding behaviours rigorously and continuously.

As a consequence, we conclude that reducing the costs of behaviour can be effective for changing specific behaviours so that they move in a sustainable direction. Because such interventions will most likely not have an influence on people's dedication to protecting the environment (their environmental attitude), these kinds of interventions should be implemented wisely. And although such measures can effectively change behaviour for the better, backfiring must be expected, that is, other unsustainable behaviours are likely to increase as a consequence.

Effectively promoting individual sustainability involves finding ways to increase people's environmental attitude so that people are motivated to act sustainably in all aspects and areas of their lives. From a practical perspective, this means that addressing secondary benefits, such as saving money, as a prime target of interventions is not advisable. Only when people take action to meet a personal sustainability goal can they be expected to implement a more sustainable lifestyle and reduce their overall energy and resource consumption.

Supportive conditions for a sustainable society

In order for Western societies to reduce their energy and resource consumption to sustainable levels, individual behaviour – and even more so, the motivation behind behaviour – matters. As long as sustainability is not a personal goal, people are unlikely to forego consumption behaviours that increase comfort, fun, status, or wealth (see Otto *et al.* 2014). However, if people hold sustainability in high regard and truly aim for environmental protection – a motivation that reveals itself in a person's environmental attitude – they will accept sacrifices and inconveniences. In other words, only with sufficient dedication to sustainability will a person seize opportunities to act sustainably and implement a sustainable lifestyle. Without committed people, there are no sustainable societies. Thus, it stands to reason that societies need to actively foster people's commitment as well.

We have argued in this chapter that the sociocultural conditions – that is, the contexts in which people make their daily decisions and in which they lead their lives – define the behavioural options that are available and determine how difficult these options are in comparison with alternatives. Some examples of these behavioural options are the type and quality of a city's public transportation system, the proportion of renewable energy in the grid, the range of organic produce available in grocery stores, and tax regulations that favour environmentally friendly products and services. All these factors and many more are relevant when people make decisions about committing to green electricity tariffs, public transportation, or a strictly organic diet, to name just a few. Sociocultural conditions are of utmost importance for the sustainable performance of individuals. If these conditions are supportive, they facilitate sustainable behaviours and sustainable lifestyles.

Such supportive conditions are not God-given facts, but they can be enforced by individuals in a bottom-up manner. Through voting, civic engagement, or choice of profession, individuals can help shape the sociocultural conditions in which they live. Favourable conditions for sustainable behaviour will enable more people to implement sustainable lifestyles. For favourable conditions to occur, it is necessary for people to be committed to sustainability because it is the people in democratic societies who determine the rigour with which policymakers implement measures to promote sustainability. In other words, both favourable conditions and favourable attitudes are needed to jointly and reciprocally create sustainable societies.

Acknowledgements

This work was supported by the German Federal Ministry of Education and Research (BMBF) as part of the Kopernikus Project ENavi [grant 03SFK4Q0]. We wish to thank Jane Zagorski for her language support and the editors of this book for their helpful comments on earlier versions of this chapter.

Notes

1 Sustainability means preserving the earth's habitability for humankind by respecting its planetary boundaries (Rockström *et al.* 2009). Climate change is a planetary boundary that requires the concentration of greenhouse gases to remain below a certain level (IPCC 2014). Planetary boundaries are affected by human energy and resource consumption, which is a result of people's lifestyles or a composite of their activities.
2 Despite Sweden's commitment to a transition to renewable energy, approximately one-third of Sweden's electricity to date is produced by nuclear power (Swedish Energy Agency 2018). While this is commonly not regarded as a solution for sustainable energy systems (German Federal Environmental Protection Agency [UBA] 2013), it does serve as an example of man-made conditions.

References

Abrahamse, W., Steg, L., Vlek, C., and Rothengatter, T. (2005) A review of intervention studies aimed at household energy conservation. *Journal of Environmental Psychology*, 25, 273–291.
Abou-Zeid, M., Witter, R., Bierlaire, M., Kaufmann, V., and Ben-Akiva, M. (2012) Happiness and travel mode switching: Findings from a Swiss public transportation experiment. *Transport Policy*, 19, 93–104.

Arnold, O. (2017) Verhalten als kompensatorische Funktion von Einstellung und Verhaltenskosten: Die Person-Situation-Interaktion im Rahmen des Campbell-Paradigmas (Behavior as a compensatory function of attitude and behavioral costs: the person-situation interaction within the Campbell Paradigm). PhD thesis, Otto von Guericke University Magdeburg, Germany.

Arnold, O. and Kaiser, F.G. (2018) Understanding the foot-in-the-door effect as a pseudo-effect from the perspective of the Campbell paradigm. *International Journal of Psychology*, 53, 157–165.

Arnold, O., Kibbe, A., Hartig, T., and Kaiser, F.G. (2018) Capturing the environmental impact of individual lifestyles: Evidence of the criterion validity of the general ecological behavior scale. *Environment and Behavior*, 50, 350–372.

Baca-Motes, K., Brown, A., Gneezy, A., Keenan, E.A. and Nelson, L.D. (2012) Commitment and behavior change: evidence from the field. *Journal of Consumer Research*, 39, 1070–1084.

Bamberg, S. (2006) Is a residential relocation a good opportunity to change people's travel behavior? Results from a theory-driven intervention study. *Environment and Behavior*, 38, 820–840.

Best, H. and Kneip, T. (2011) The impact of attitudes and behavioral costs on environmental behavior: a natural experiment on household waste recycling. *Social Science Research*, 40, 917–930.

Bilharz, M. (2008) *'Key Points' nachhaltigen Konsums. Ein strukturpolitisch fundierter Strategieansatz für die Nachhaltigkeitskommunikation im Kontext aktivierender Verbraucherpolitik (The key points of sustainable consumption. A policy structure-based approach for a sustainability-centred communication strategy within the context of an activating consumer policy)*. Marburg, Metropolis.

Brügger, A., Kaiser, F.G., and Roczen, N. (2011) One for all? Connectedness to nature, inclusion of nature, environmental identity, and implicit association with nature. *European Psychologist*, 16, 324–333.

Byrka, K., Kaiser, F.G., and Olko, J. (2017) Understanding the acceptance of nature-preservation-related restrictions as the result of the compensatory effects of environmental attitude and behavioral costs. *Environment and Behavior*, 49, 487–508.

Clayton, S. (2003) Environmental identity: a conceptual and an operational definition. In S. Clayton and S. Opotow (eds), *Identity and the Natural Environment: The Psychological Significance of Nature*. Cambridge, MA: MIT Press.

Darby, S. (2006) The effectiveness of feedback on energy consumption: a review for DEFRA of the literature on metering, billing and direct displays. Available at:www.eci.ox.ac.uk/research/energy/downloads/smart-metering-report.pdf, (accessed 9 August 2018).

Díaz-Siefer, P., Neaman, A., Salgado, E., Celis-Diez, J.L. and Otto, S. (2015) Human-environment system knowledge: a correlate of pro-environmental behavior. *Sustainability*, 7, 15510–15526.

Eurostat (2014) Greenhouse gas emissions per capita in tonnes of CO_2 equivalent per capita. Available at: http://ec.europa.eu/eurostat/tgm/table.do?tab=tableandinit=1andlanguage=enandpcode=t2020_rd300andplugin=1 (accessed 9 May 2018).

German Federal Environmental Protection Agency [UBA] (2013) Costs of meeting international climate targets without nuclear power. Available at: www.umweltbundesamt.de/sites/default/files/medien/378/publikationen/climate_change_11_2015_knoche_nuclear_phase_out_komplett.pdf (accessed 7 August 2018).

German Federal Environmental Protection Agency [UBA] (2018) Energieverbrauch privater Haushalte [Energy consumption of private households]. Available at: www.

umweltbundesamt.de/daten/private-haushalte-konsum/wohnen/energieverbrauch-privater-haushalte (accessed 24 January 2018).

Gölz, S. (2017) Does feedback usage lead to electricity savings? Analysis of goals for usage, feedback seeking, and consumption behavior. *Energy Efficiency*, 10, 1453–1473.

Grunwald, A. (2010) Wider die Privatisierung der Nachhaltigkeit–Warum ökologisch korrekter Konsum die Umwelt nicht retten kann [Against Privatisation of Sustainability – Why Consuming Ecologically Correct Products Will Not Save the Environment]. *GAIA – Ecological Perspectives for Science and Society*, 19, 178–182.

Hamann, K.R., Reese, G., Seewald, D., and Loeschinger, D.C. (2015) Affixing the theory of normative conduct (to your mailbox): injunctive and descriptive norms as predictors of anti-ads sticker use. *Journal of Environmental Psychology*, 44, 1–9.

Henn, L., Otto, S., and Kaiser, F.G. (2019a) Positive spillover: the result of attitude change. Manuscript in preparation.

Henn, L., Taube, O., and Kaiser, F.G. (2019b) The role of environmental attitude in the efficacy of smart-meter-based feedback interventions. *Journal of Environmental Psychology*, 63, 74–81.

Hinds, J. and Sparks, P. (2008) Engaging with the natural environment: the role of affective connection and identity. *Journal of Environmental Psychology*, 28, 109–120.

Hoekstra, A.Y. and Wiedmann, T.O. (2014) Humanity's unsustainable environmental footprint. *Science*, 344(6188), 1114–1117.

Hoppmann, J., Huenteler, J., and Girod, B. (2014) Compulsive policy-making—the evolution of the German feed-in tariff system for solar photovoltaic power. *Research Policy*, 43, 1422–1441.

IPCC (2014) Climate Change 2014: Synthesis Report. Contribution of Working Groups I, II and III to the Fifth Assessment Report of the Intergovernmental Panel on Climate Change [Core Writing Team, R.K. Pachauri and L.A. Meyer (eds)]. Geneva: IPCC.

Ivanova, D., Vita, G., Steen-Olsen, K., Stadler, K., Melo, P.C., Wood, R. and Hertwich, E.G. (2017) Mapping the carbon footprint of EU regions. *Environmental Research Letters*, 12(5), 054013.

Kaiser, F.G., Arnold, O., and Otto, S. (2014a) Attitudes and defaults save lives and protect the environment jointly and compensatorily: understanding the behavioral efficacy of nudges and other structural interventions. *Behavioral Sciences*, 4, 202–212.

Kaiser, F.G., Brügger, A., Hartig, T., Bogner, F.X., and Gutscher, H. (2014b) Appreciation of nature and appreciation of environmental protection: how stable are these attitudes and which comes first? *European Review of Applied Psychology/Revue Européenne de Psychologie Appliquée*, 64, 269–277.

Kaiser, F.G., Byrka, K., and Hartig, T. (2010) Reviving Campbell's paradigm for attitude research. *Personality and Social Psychology Review*, 14, 351–367.

Kaiser, F.G. and Byrka, K. (2015) The Campbell paradigm as a conceptual alternative to the expectation of hypocrisy in contemporary attitude research. *The Journal of Social Psychology*, 155, 12–29.

Kaiser, F.G., Hartig, T., Brügger, A., and Duvier, C. (2013) Environmental protection and nature as distinct attitudinal objects: an application of the Campbell paradigm. *Environment and Behavior*, 45, 369–398.

Kaiser, F.G., Merten, M., and Wetzel, E. (2018) How do we know we are measuring environmental attitude? Specific objectivity as the formal validation criterion for measures of latent attributes. *Journal of Environmental Psychology*, 55, 139–146.

Kaiser, F.G. and Wilson, M. (2004) Goal-directed conservation behavior: the specific composition of a general performance. *Personality and Individual Differences*, 36, 1531–1544.

Kaiser, F.G. and Wilson, M. (in press) The Campbell Paradigm as a behavior-predictive reinterpretation of the classical tripartite model of attitudes. *European Psychologist*.

Kibbe, A. (2017) Intrinsische Umweltmotivation. Selbstbestimmungstheorie und Campbell-Paradigma im Vergleich (Intrinsic environmental motivation. A comparison of self-determination theory and Campbell Paradigm). PhD thesis, Otto-von-Guericke University Magdeburg, Germany.

Klöckner, C.A., Nayum, A., and Mehmetoglu, M. (2013) Positive and negative spillover effects from electric car purchase to car use. *Transportation Research Part D: Transport and Environment*, 21, 32–38.

Klöckner, C.A. and Ofstad, S.P. (2017) Tailored information helps people progress towards reducing their beef consumption. *Journal of Environmental Psychology*, 50, 24–36.

Lokhorst, A.M., Werner, C., Staats, H., van Dijk, E., and Gale, J.L. (2013) Commitment and behavior change: a meta-analysis and critical review of commitment-making strategies in environmental research. *Environment and Behavior*, 45, 3–34.

Maki, A., Burns, R.J., Ha, L., and Rothman, A.J. (2016) Paying people to protect the environment: a meta-analysis of financial incentive interventions to promote proenvironmental behaviors. *Journal of Environmental Psychology*, 47, 242–255.

Mayer, F.S. and Frantz, C.M. (2004) The connectedness to nature scale: a measure of individuals' feeling in community with nature. *Journal of Environmental Psychology*, 24, 503–515.

McNeill, J. (2000) *Something New Under the Sun: An Environmental History of the 20th Century*. New York: Norton.

Nilsson, A., Bergstad, C.J., Thuvander, L., Andersson, D., Andersson, K., and Meiling, P. (2014) Effects of continuous feedback on households' electricity consumption: potentials and barriers. *Applied Energy*, 122, 17–23.

Nisbet, E.K., Zelenski, J.M., and Murphy, S.A. (2009) The nature relatedness scale. Linking individuals' connection with nature to environmental concern and behavior. *Environment and Behavior*, 41, 715–740.

Osbaldiston, R. and Schott, J.P. (2012) Environmental sustainability and behavioral science: Meta-analysis of proenvironmental behavior experiments. *Environment and Behavior*, 44, 257–299.

Otto, S. and Kaiser, F. G. (2014) Ecological behavior across the lifespan: why environmentalism increases as people grow older. *Journal of Environmental Psychology*, 40, 331–338.

Otto, S., Kaiser, F.G., and Arnold, O. (2014) The critical challenge of climate change for psychology: preventing rebound and promoting more individual irrationality. *European Psychologist*, 19, 96–106.

Rockström, J., Steffen, W., Noone, K., Persson, Å., Chapin III, F.S., Lambin, E.F., ..., and Foley, J. (2009) A safe operating space for humanity. *Nature*, 461(7263), 472–475.

Roczen, N., Kaiser, F.G., Bogner, F.X., and Wilson M. (2014) A competence model for environmental education. *Environment and Behavior*, 46, 972–992.

Scarborough, P., Appleby, P.N., Mizdrak, A., Briggs, A.D., Travis, R.C., Bradbury, K.E., and Key, T.J. (2014) Dietary greenhouse gas emissions of meat-eaters, fish-eaters, vegetarians and vegans in the UK. *Climatic Change*, 125, 179–192.

Schultz, P.W. (2014) Strategies for promoting proenvironmental behavior: lots of tools but few instructions. *European Psychologist*, 19, 107–117.

Steffen, W., Persson, Å., Deutsch, L., Zalasiewicz, J., Williams, M., Richardson, K., ..., and Molina, M. (2011) The Anthropocene: from global change to planetary stewardship. *AMBIO: A Journal of the Human Environment*, 40, 739–761.

Steg, L. and Vlek, C. (2009) Encouraging pro-environmental behaviour: an integrative review and research agenda. *Journal of Environmental Psychology*, 29, 309–317.

Stern, P.C. (2000) New environmental theories: toward a coherent theory of environmentally significant behavior. *Journal of Social Issues*, 56, 407–424.

Swedish Energy Agency (2018) Statistics: energy in Sweden 2018. Available at: www. energimyndigheten.se/en/facts-and-figures/statistics/ (accessed 7 August 2018).

Taube, O., Kibbe, A., Vetter, M., Adler, M., and Kaiser, F.G. (2018) Applying the Campbell Paradigm to sustainable travel behavior: compensatory effects of environmental attitude and the transportation environment. *Transportation Research Part F: Traffic Psychology and Behaviour*, 56, 392–407.

Taube, O. and Vetter, M. (2019) How green defaults promote environmentally-friendly decisions: Attitude-conditional default acceptance but attitude-unconditional effects on actual choices. Manuscript submitted for publication.

Thaler, R. and Sunstein, C. (2008) *Nudge: The Gentle Power of Choice Architecture*. New Haven, CT: Yale University Press.

Thøgersen, J. (2005) How may consumer policy empower consumers for sustainable lifestyles? *Journal of Consumer Policy*, 28, 143–177.

Thøgersen, J. (2014) Unsustainable consumption: basic causes and implications for policy. *European Psychologist*, 19, 84–95.

Thomas, C. and Sharp, V. (2013) Understanding the normalisation of recycling behaviour and its implications for other pro-environmental behaviours: a review of social norms and recycling. *Resources, Conservation and Recycling*, 79, 11–20.

Toft, M.B., Schuitema, G., and Thøgersen, J. (2014) The importance of framing for consumer acceptance of the Smart Grid: a comparative study of Denmark, Norway and Switzerland. *Energy Research and Social Science*, 3, 113–123.

Truelove, H.B., Carrico, A.R., Weber, E.U., Raimi, K.T., and Vandenbergh, M.P. (2014) Positive and negative spillover of pro-environmental behavior: an integrative review and theoretical framework. *Global Environmental Change*, 29, 127–138.

United Nations (2015) Paris Agreement. Available at: https://unfccc.int/sites/default/ files/english_paris_agreement.pdf (accessed 9 May 2018).

Urban, J. and Ščasný, M. (2016) Structure of domestic energy saving: how many dimensions? *Environment and Behavior*, 48, 454–481.

Wang, T.H. and Katzev, R.D. (1990) Group commitment and resource conservation: two field experiments on promoting recycling. *Journal of Applied Social Psychology*, 20, 265–275.

Webb, T.L., Benn, Y., and Chang, B.P. (2014) Antecedents and consequences of monitoring domestic electricity consumption. *Journal of Environmental Psychology*, 40, 228–238.

Yu, B., Zhang, J., and Fujiwara, A. (2013) Evaluating the direct and indirect rebound effects in household energy consumption behavior: a case study of Beijing. *Energy Policy*, 57, 441–453.

3 'It has to be reasonable'

Pragmatic ways of living sustainably in Danish eco-communities

Anette Høite Hansen

Introduction

Most, if not all, studies of and publications on eco-communities recognise the environmentally, economically, and socially sustainable advantages and opportunities of eco-communities. Such studies describe the sustainable ideals and visions articulated by residents of eco-communities about how to build houses, what kind of food to eat, and what degree of social interaction to have with neighbours (Marckmann 2009; Marckmann *et al.* 2012; Lockyer and Veteto 2013; Litfin 2014; LØS 2016; Miller 2018). These ideals and visions can be seen as the residents' justification for their decision to criticise and also distance themselves from life as lived in mainstream society. Nevertheless, although residents of eco-communities are often perceived to have adopted a high degree of sustainable behaviour, they do not necessarily have the exact same self-image about the choices they have made. According to the chairperson of the Danish Association of Eco-Communities, LØS,[1] the term 'eco-community' is understood by some residents as sounding too self-righteous because residents in fact believe that they are not doing 'as much as they could' and because they sometimes make compromises with their sustainable ideals (Bisgaard 2018).

I agree with the analysis of eco-communities as being suitable homes for people who have a desire to live in an arena where sustainable ideals and practices are often easier to fulfil compared with conventional individual housing. However, my analysis of eco-communities in this chapter discusses the reasonableness of these ideals and examines how sustainable behaviour is understood, practised, and navigated by the residents of eco-communities.

The 1987 Brundtland Report, 'Our Common Future', defined sustainable development as: '[…] development that meets the needs of the present without compromising the ability of future generations to meet their own needs' (Brundtland and World Commission on Environment and Development 1987). Since that report, there has been a lively discussion about what the concept of sustainability is and what it should imply. In 2015, sustainable development gained an increased focus with the UN's 2030 agenda, entitled 'Sustainable Development Goals' (SDGs). The SDGs consist of 17 different goals that address a variety of sustainability concerns such as sustainable energy, access to clean water, and a decrease in inequality (United Nations 2019). Hence, both the Brundtland Report and the SDGs aim to ensure sustainable *development* around the world. However, many climate change scholars, social scientists, and other sustainability experts argue that development that meets the needs of the present is not sustainable; instead, a whole transition away from how we live today is necessary to prevent any more environmental, social, and economic damage caused by climate change than is already predicted (Hopkins 2008; IPCC 2018). With residents of eco-communities as the pivotal point of this chapter, I focus on a setting where this (necessary) sustainable transition, expressed in smaller communities, is to some degree both idealised and practised.

On my first day of fieldwork in an eco-community I talked to a 39-year old female resident who had just returned from a summer holiday with her husband and their three children. While on holiday, they had visited an eco-community in Slovenia that the resident said had a relatively strict way of interpreting sustainable living. For instance, members of this eco-community do not drink coffee, presumably because they do not grow the coffee beans themselves. Hence, the resident felt quite relieved when she and her family left this eco-community after a few days, and she could get herself a cup of coffee. The resident concluded her story by making a telling remark: 'It has to be reasonable'. She continued, saying that she could not live in an eco-community with such strict principles about sustainability and finds the sustainability principles of the Danish eco-community she lives in exactly that: *reasonable*. This story is one of many times that I have heard the phrase 'it has to be reasonable' in terms of principles of sustainability in Danish eco-communities. For instance, the phrase is expressed when residents debate whether their physical and social structures could and should be revised in order to enable either more or less sustainable behaviour or when they consider whether to spend time in their private households or in the communal space.

I describe in this chapter how the everyday lives in Danish eco-communities are organised and how sustainability is conceived through the *pragmatic approach* that the statement 'it has to be reasonable' illustrates. I argue that the liveability of an eco-community depends on the residents' ability to be pragmatic about the extent of their sustainable behaviour as well as their taking part in the social aspects of living in a community. 'It has to be reasonable' captures the spirit of an often-used though never clearly defined approach to living in an eco-community and practising sustainable behaviour, which illustrates how the capacity of being both sustainable and engaged in an eco-community is dynamic and ever-changing. This chapter addresses the importance of striking a balance between sustainable ideals, collective engagement, and individual considerations in Danish eco-communities.

The field: Danish eco-communities

An eco-community is a type of dwelling that covers many different ways of organising homes and housing facilities.[2] In general, living in an eco-community implies that the residents live in homes situated next to each other in a defined community, which is established as an association with some sort of environmentally sustainability as well as a collective focus on its set of values (Reinholdt 1997; Marckmann 2009; Marckmann *et al.* 2012; Lockyer and Veteto 2013; Litfin 2014; LØS 2016; Miller 2018). Eco-communities vary in size, design, profile, and geography. While some are located in Danish towns or larger cities, others are situated in the countryside near smaller villages. Some are built as terraced houses and others are unique single houses designed by the owner. Common to a majority of eco-communities is the fact that they have a communal building as the centre of the community activities, such as dinner, meetings and social gatherings. There has been a remarkable increase in interest in eco-communities and other kinds of co-housing in Denmark in recent years (Udlændinge-, Integrations- og Boligministeriet [The Ministry of Immigration, Integration and Housing] 2016; LØS 2016). There has also been an increased interest in initiating new eco-community projects. There are currently around 50 eco-communities in Denmark, while another 30 are in the conceptual phase (LØS 2018). The number of residents varies from around 60 up to almost 300 in the largest eco-communities. The larger eco-communities are often divided into smaller groups to facilitate certain areas of responsibility and decision-making processes.

Over a period of nine months in 2017–2018, I collected anthropological fieldwork data centred on three different eco-communities; each community expressed a different focus on sustainability in their set of values. Whereas one eco-community experimented with self-sufficiency in terms of food, another prioritised being self-sufficient with energy, while the third initiated local socially and environmentally sustainable businesses. In addition to my fieldwork studies in these eco-communities, I have participated in conferences, seminars, excursions, and courses in the Danish as well as the European eco-community and eco-village movements (LØS and GEN-Europe[3]). The in-depth research in the three different eco-communities consisted of longer stays, when I lived in private homes and guesthouses or visited on a daily basis. I participated in both the preparation and eating of communal dinners, community work such as gardening or cleaning waste storage rooms, communal and working group meetings, social gatherings, and various celebrations. I also participated in dozens of spontaneous everyday activities and situations that were inevitably woven in and around these duties. Furthermore, I visited the private family homes, where I drank coffee and ate lunch with the informants and played games with their children. It was in the private homes that most of my semi-structured interviews took place. Thus, I have both lived and studied the everyday life[4] of residents in Danish eco-communities, which provided me with valuable insights into how people attempt to merge their interest in sustainability, their engagement in community life, and their everyday family life.

Both the names of eco-communities I studied and the names of their residents are anonymised here. My research has primarily, though not exclusively, focused on families with children. The empirical material that follows describes in particular the advantages as well as the compromises of sustainable living from the perspective of families. Had the focus been on young couples, seniors or single people, the analysis would undoubtedly be different. However, from my insight gained through interviews and conversations with a broad group of residents in the eco-communities, I argue that the importance of a *pragmatic approach* to sustainable living is paramount for a majority of the residents. Nevertheless, these pragmatic views naturally take on different forms depending on the living situations of the residents.

Theoretical approach

A renewed interest in the concept of pragmatism has occurred in the fields of anthropology and other social sciences in recent decades (Whyte 1997; Brinkmann 2006; Gausset 2010; Goldman 2012). From an anthropological viewpoint, Susan R. Whyte argues that pragmatism is '[...] particularly applicable in relation to understanding problems, handling of problems and the search for problem-solving' (Mogensen and Whyte 2004, 47, my translation). Following Whyte, the concept of pragmatism is applicable in the current study of eco-communities since I analyse how residents cope with and react to challenges in their communal everyday lives. As a philosophical school of thought, *pragmatism* is seen as a way of understanding things in relation to their consequences. The American philosopher William James (1981 [1907]) operates with what he calls *pragmatic methods*:

> No particular results then, so far, but only an attitude of orientation, is what the pragmatic method means. The attitude of looking away from first things, principles, 'categories', supposed necessities, and of looking towards last things, fruits, consequences, facts.
>
> (James 1981 [1907], 29)

This chapter makes use of James's concept of pragmatism to better understand the structural and collective sustainable dynamics in Danish eco-communities. In analysing how residents of Danish eco-communities have come to focus more on the *consequences* of their sustainable practices rather than their original principles, ideals, and visions of living in an eco-community, I distinguish between what I call *structural pragmatism* and *collective pragmatism*. I define *structural pragmatism* as the way residents of eco-communities react upon consequences by negotiating and changing the physical and social structures that enable a sustainable behaviour. *Collective pragmatism* is understood as the way residents react to consequences by balancing between engaging in collective activities and life in their private households. Through empirical examples, I show how processes of structural pragmatism and collective pragmatism are mutually significant as well as crucial for the very liveability of the eco-communities.

Another approach relevant to understanding sustainable behaviour in Danish eco-communities is *practice theory*, since it is through the residents' *sustainable practices* that a pragmatic approach in the eco-communities appears. What I call *sustainable practices* are practices aimed at minimising human impact on the environment. In the Danish eco-communities that I have studied, these practices take two major forms: physical and social structures that work as frameworks for the resident's daily actions. The physical structures are, for instance, communal buildings and gardens, washing rooms, solar panels, sewage plants, and other facilities that residents share, but also the individual houses that are constructed from sustainable principles and designed to promote sustainable behaviour. The social structures are, for instance, the communal meetings, the different communal work responsibilities, and the communal dinners as well as other social gatherings and daily encounters among the residents. According to Elizabeth Shove and Nicola Spurling (2014, 1), '[…] social theories of practice provide an important intellectual resource for understanding and perhaps establishing social, institutional and infrastructural conditions in which much less resource intensive ways of life might take hold'. Thus, the present analysis on sustainable living in Danish eco-communities – generally understood as 'less resource-intensive ways of lives' – orients towards *practice theory*. However, as Shove and Spurling argue, there is no one exclusive theory of practice: there is '[…] no such thing as "*a*" practice approach […]' (Shove and Spurling 2014, 3, emphasis added). Following from this approach, 'practice' is here understood as an empirical tool to emphasise the relevance of the pragmatic method rather than as an analytical framework in itself. Furthermore, focusing on practice is a necessary tool to describe the collectively oriented sustainable living of residents in eco-communities who interact with one another in their daily lives, since 'practices […] cannot be conceived as a set of individual actions, but […] are essentially modes of social relations, of mutual action' (Taylor in Shove *et al.* 2012, 5).

Pragmatic ways of living sustainably

The three different eco-communities in which I conducted fieldwork have gone through similar processes of development since they were organised and became operational. Even before a piece of land was chosen upon which to build these eco-communities, the future residents held various meetings and decided on their common sets of values. These values formed the basis of what the physical and social structures of the everyday lives in the future eco-communities should look like. This process took several years from when the initiators began to meet until the first sod was turned in the fields of the eco-communities; despite the fact that the three eco-communities were initiated 26, 15 and six years ago, respectively, the residents say that they are in a continuous state of formation. These formation processes are characterised by expansion, that is, the building of more houses and the optimisation of the communal surroundings such as new communal buildings being constructed or new trees being planted. However, it has become clear that despite all the planning and discussions that have occurred through the life of these communities that the social formation of collective life has turned out to be more demanding than the initiators had imagined.

For the residents who have arrived most recently, the motivation for moving into an eco-community tends towards a focus on the social aspects of 'living in community'[5] rather than a strong ideal of living with environmentally sustainable values. This motivation is not an expression of the residents caring less about the environment, nor does it mean that their practices in the eco-community are less sustainable than those of the earliest residents. Instead, this motivation indicates that the sustainable structures of the eco-community have already been established, which means that the new residents do not need to take environmental considerations into account in the same way as the earliest residents did when they created their values statement and started establishing the eco-community. However, this does not mean that the sustainable structures are not continuously revised as the opinions and values of the residents change over time.

When it comes to the founding ideals, all three eco-communities have gone through rather serious reassessments, which in some cases have led to some of the earliest residents leaving the community. One eco-community had an initial vision of being 'as self-sustaining as possible', but this approach occasionally caused conflicts among residents since it turned out to be difficult to define when the eco-community was adequately self-sustaining. Some residents intended to reach a higher level of environmentally sustainable principles and were comfortable with the effort required to do so. But others found that their having to grow and produce much more food than what was already the case would require spending too much time on what was intended to be a hobby farm while attending to other duties such as work, private family time, etc. Thus, the initial focus in this eco-community turned out to put too much strain on the strong ideals of living sustainably and too little on other important aspects of the resident's lives. This is an example that residents describe as the moment when they realised that 'it has to be reasonable' to live in an eco-community and that it is important to be pragmatic when deciding what the physical and social structures of their everyday lives and the necessity of collective engagement should be like. Hence, being pragmatic is an emic understanding, which illustrates how a balanced and reasonable approach to a sustainable everyday life is desired.

Concurrently with the emic reasonable approach, I make use in the following analysis of James's understanding of pragmatism by showing how the residents often turn their focus away from initial principles and consider the consequences of these principles. I argue that the residents of the eco-communities continuously put themselves through processes where they balance the consequences they face in their attempts to practice sustainable behaviour and the sustainable principles that they initially had intended to follow. These compromises happen continuously as structures and systems are constantly revised. Subsequently, the residents often return to narratives of more radical experiences where the cohesion of the community overruled the initial sustainable ideals. For instance, in one eco-community, the residents sought external help from therapists to work on their disagreements and the bad atmosphere that had become prevalent. This can be seen as a reminder that if compromises are not made, internal conflicts among residents will occur. Thus, the experience of serious challenges that occur when acting on principles rather than considering consequences has resulted in a renewed awareness of the need to factor in consequences in the eco-communities' decision-making processes as well as in everyday life practices in general.

How sustainable the everyday life practices should be is a theme that is generally and continually debated in all the eco-communities I have studied. For example, one eco-community I studied debated about whether the communal dinner should be vegetarian or not. Some residents thought that eating vegetarian at the communal dinner was a natural aspect of living in an eco-community, but others were frustrated about someone else deciding whether they should eat meat or not. It became obvious that while the residents were busy planning the physical structures of this eco-community – the architectural design of the houses, the renewable energy installations, and so forth – they had underestimated the crucial importance of debating the issues of social structures and collective engagement. As a 67-year old male resident put it:

> We have realised that being self-sufficient with energy simply cannot be the principal element – a social focus on community and the consideration to people's different opinions has to be taken into account as well.

After the disagreement about vegetarian food had faded, several residents actually voluntarily chose the vegetarian alternative at the communal dinner in this eco-community. Thus, the disagreement among the residents turned out to be less about sustainable practices and more about having to make compromises and the right to decide for oneself. Again, this shows how a *pragmatic method* is applied in the revision of ideals in order to protect social cohesion.

In the following sections, I describe in detail how the pragmatic method, understood as *structural pragmatism* and *collective pragmatism* respectively, unfolds in these Danish eco-communities.

Structural pragmatism

A 40-year-old female resident and mother of three describes the structural composition of her eco-community:

> In the eco-community, we have a lot of structures. Up here in the farm building [which functions as the communal building in this eco-community], every day, all year round, except around three–four days around Christmas, food is cooked for everyone who feels like joining the communal dinner. And this communal dinner is a structure, you see. And it has a lot to do with sustainability. [...] We also have shared food here in the farm building. Not just for dinner. When we are preparing breakfast and

lunch at home, we go here and pick up the produce that we need. In fact, this means that we have a big food box, a shared food economy – we all pay the same amount of money every month, and then food is either produced by ourselves or groceries are bought through our wholesale system. It is one of the responsibilities of the communal work tasks to order groceries because all of us also have work responsibilities in the community, and we are committed to work four hours a week for the community. We have some principles about the food that we buy or that we produce – this is also a structure. Then there is another thing about the produce we buy: we have some principles saying that they should not be cultivated in a heated green-house. Additionally, the transport of the groceries should not be too long. We could question whether this is the right thing to do, but we have agreed that groceries should be able to be transported by truck to our community. Hence, the groceries might be from Southern Europe, but they should not come from Africa or be transported by aeroplane. You see, this is a carbon-emission issue, one could say. But then there are other things that we are not aware of. For example, we sometimes buy avocados, and one could ask: 'Avocados, is that sustainable, isn't a lot of water used in watering them?' Yes, there is. So there are also structures in the food system that are not that sustainable. But that is the thing – when you try to create a framework for how we do things then it will, at times, also fall through.

In this statement, a resident narrates how the eco-community in question is structured and designed to promote an environmentally sustainable food consumption. Her story reflects that sustainability is understood by this eco-community as something that has to do with animals that are bred at home and food that is produced as locally as possible and transported as little as possible. However, these principles create dilemmas at times: they can be difficult to maintain, as it was the case with the issue of eating avocados. The organising of the food system is continuously debated in this eco-community and changes from time to time when compromises among residents require it.

When this eco-community first started, all produce was shared in the food system, including what the residents call 'luxury goods' such as cheese, raisins, cereals, coffee, and the like. However, at one point, the eco-community realised that paying for these groceries jointly created some greed and a 'tragedy of the commons' (Hardin 1968), that is, the residents tended to act according to their own self-interest and not for the common good of their fellow eco-community residents, which was the purpose of having a shared food system in the first place. This problem could be avoided if the residents had to pay for their groceries individually. As a result, the wholesale system in this community is now organised as a 'basic produce system' for universally shared produce that is jointly paid for in advance while the 'luxury goods' are bought privately. This is a clear, straightforward illustration of *structural pragmatism*: not every resident agrees with the division between basic and luxury goods, but the residents have agreed that this is the best solution for the eco-community in general. Thus, the re-organising of the food system is an example of how the residents structurally and pragmatically react to consequences (that too many luxury goods are consumed when paid jointly) rather than stick to their principle ideal of sharing the cost of all the food consumed by the community.

Other structures are often up for debate in eco-communities. In the following statements, two female residents, aged 40 and 52 years and mothers of three and two children respectively, discuss how some vegetarians compromise with their actual desires and ideals on living in an eco-community without breeding animals:

> The thing is, we obviously have some vegetarians. We are not all carnivores, you see. And through their monetary contribution to the food system the vegetarians pay for meat – we do not have such a differentiated food contribution – they all pay the same food contribution like I do, even though we know that meat is costly. It especially costs in terms of the use of land space. We try to produce our fodder ourselves, but once in a while we buy fodder from external sources. And that is simply something they have accepted, these vegetarians. But it is one of these discussions that comes up again and again. If one could say that sometimes the individual person can be more ambitious than the community, then this is occasionally the case for vegetarians. Because those of us that eat meat, we would actually have liked to have a few more pasture-fed cows during the summer, because self-produced beef is just very, very delicious in comparison with having to go and get it in the supermarket if I need beef for cooking at home. But in this case, the vegetarians have simply said 'we will not take part in this'.
>
> [...].
>
> So one could say: The vegetarians would like it if there was no meat at all in the village, so the structures we have also entails that vegetarians and vegans are not fully capable of living the way they actually desire. So, it is an illustration of the fact that structures can make it harder for someone to get what they want.

These residents indicate that both carnivores and vegetarians in eco-communities compromise their desire to breed more or fewer cattle. As in the case of the wholesale system, this question about cattle breeding is an ongoing debate at communal meetings and in everyday conversations among residents. On the one hand, the breeding of animals interferes with an ideal of environmentally sustainable behaviour because of the emission of greenhouse gases that results from breeding, and the land needed to produce fodder. On the other hand, having animals is also a natural part of living in the countryside, and residents argue that the eco-communities' way of breeding provides a more desirable production of the meat compared with meat bought in the local supermarket. Hence, the compromise on the issue of cattle breeding is another example of *structural pragmatism*: since having no animals in the eco-community is a sustainable ideal that interferes with some residents' desires, taking this consequence into account has resulted in a compromise of breeding just a few cattle as part of the community structure. At the same time, this compromise illustrates how the structures enable carnivores to eat less meat than they would if they lived by themselves and decided the number of bred animals to breed on their own. Accordingly, embedded structures of the eco-community help the residents in their effort to eat in a sustainable way, which is otherwise experienced as a complicated matter with a lot of decisions and reflections to be made.

A 40-year-old female resident and mother of three appreciates the benefit of the shared decision-making of her eco-community's food system:

> If one is to emphasise another thing about this food system, then one could say that having these structures prevents me from having to stand in the supermarket, a little hungry, on a Thursday afternoon with three kids and figuring out what we should eat for dinner. Should it be meat or chicken or chickpeas, or what should it be? In the eco-community, someone has made that decision for me. Someone has made the choice for me. And the thing is, if one would like to eat sustainably, it is extremely complex. It is extremely complex to get acquainted with conditions of production, and, well, how much water was used for that, and how far was it transported; is it transported by aeroplane or is it transported by ship? And how about this cow, has it eaten anything that comes from South America or has it grazed in a field? There are so many things that one has to consider all the time. So the fact is that we have a framework. And that someone has made the decisions for us [...], it just makes it extremely easy, right?

As this resident describes, the advantage of a shared food system, and of a communal dinner in particular, is not just the fact that the residents do not have to consider what food is produced sustainably and what is not. The shared food system also means that they do not need to both cook and take care of their children in the busy hours after school and work. A 37-year-old male resident and father of three children also recognises this benefit:

> That communal dinner is, simply put, just a brilliant concept! My wife and I often sit and drink coffee in the afternoon after having arrived from work and picking up kids at school and day-care institutions instead of turning to the cooking of dinner directly after coming home. Our children just run out and play with the other kids in the eco-community, while the current food group cooks the communal dinner. We only take part in the dinner preparations every fifth week. When we lived in the city in a third-floor apartment I had exactly a seven-minute bicycle ride from work to the kindergarten to figure out what to cook. Then we would rush to the supermarket, and at no later than 5 o'clock we had to start cooking. And then after dinner time, we were just stuck in the apartment. It was just pure survival.

The anthropologist Sarah Berthoû argues that '[...] pro-environmental practices should not be seen as one demarcated field, but as interlinked with other practices in everyday life' (Berthoû 2013, 53). In Berthoû's study of pro-environmental behaviour[6] among citizens in Copenhagen, she describes how her interlocutors constantly 'negotiate' their everyday life practices and consider these in relation to their idea of 'the good life', which is not always in compliance with their definition of pro-environmental behaviour, e.g. taking long hot showers or flying to go on holiday (Berthoû 2013, 60f.). Furthermore, Berthoû argues that everyday life is complex: choosing to behave pro-environmentally is often experienced as quite complicated and a great responsibility for the individual (Berthoû 2013, 58f.).

Like the citizens of Copenhagen in Berthoû's case, residents of eco-communities live multifaceted lives full of social relations with family, friends, neighbours, and colleagues as well as economic and practical obligations such as work, caring for children, and community tasks. However, what distinguishes the citizens of Copenhagen from residents of the eco-communities is precisely the fact that living 'in community' provides everyday life structures that facilitate and promote sustainable choices. Having these sustainable structures and communal routines to follow makes it easier for members of eco-communities to adopt sustainable behaviour. The citizens of Copenhagen respond to the challenges of fulfilling pro-environmental ideals by continually asserting that 'You have to live' (Berthoû 2013, 61). This phrase is comparable to saying 'it has to be reasonable', which, as noted above, residents in eco-communities often say when they refer to their community's approaches to sustainable practices, such as deciding for oneself whether to eat vegetarian or not.

Thus, both individual citizens in Copenhagen and members of eco-communities are confronted with the need to make (pragmatic) compromises with their values and pro-environmental/sustainable ideals. Nevertheless, collective structures found in eco-communities ensure that members of eco-communities end up making fewer compromises in terms of sustainable behaviour than if they were living in individual housing. The main difference between the eco-community and mainstream lifestyles is that the pragmatism of Copenhageners is negotiated by individuals according to their own conscience, while the pragmatism found in eco-communities is a question of negotiating collective structures that organise individual practices. Thus, living in an eco-community promotes a sustainable behaviour because people have to adapt to community structures.[7] Such structures are organised, built, and practised as a result of long discussions and sometimes even conflicts and are therefore the result of structural pragmatism.

As we have seen, structural pragmatism unfolds on two levels: as the product of compromises and as the cause of new compromises that are constantly being made. However, as I show in the following, the liveability of eco-communities and the sustainable practice of residents also require *collective pragmatism*, which implies that people must find the right balance between private and communal life.

Collective pragmatism

A 52-year-old female resident and mother of two describes a pivotal premise of living in an eco-community:

> We have structures concerning our own produce and the food we buy in common. But no one interferes with what happens at home. So I could buy all the meat I feel like in Brugsen or Netto [Danish supermarkets] or where I feel like and eat it at home. No one will raise their eyebrows or interfere with me. It is only when negotiating structures and collective rules that people can disagree.

In other words, the collective sustainable structures (what could be called collective 'rules') of everyday life exists in the community space, not in people's private homes. However, how to balance between collective engagement and life in the private households can be experienced as a continuing negotiation in search for the right balance, as the 40-year-old female resident and mother of three describes:

> In my family we have started to eat at home half of the time because, to put it bluntly, my children, they don't eat the food that is served up here in the communal dining room. So we have made a compromise: they like meat, right, but on the other hand, they also like…. It's cosy, it's cosy to sit and eat together as a family, and it is also cosy to be up here [at the communal dinners]. So now we are doing both. Well, and I see now, when I have to cook at home for my family, which I haven't done for the last ten years, now that I have started doing this, I don't spend the same time cooking as I do when I cook the communal dinners. And following this, I use different produce at home. I end up using a lot of meat, also because the children like it, but also tomatoes, cucumbers, and so on – produce that is easy and quick to cook for this meal. Whereas, when you eat at the communal dinner, the cook has been preparing for five–six hours, so there has been time to process the produce. And this just makes it easier to make use of kale for example – a local produce, the produce that we have here in Denmark, especially in the winter time, if we should try to be self-sustaining.

Eating at home versus in the community building is not just about what type of food the residents eat but also about the need to spend some time alone with their families. This resident admits that when cooking at home, she resorts to what she calls 'produce that is easy and quick to cook', such as meat and vegetables grown in greenhouses, whereas, had she eaten at the communal dinner the same day, she would have eaten the more environmentally sustainable kale. Nevertheless, this communal consideration of sustainable eating is something she compromises with in order to give priority to her private family life.

Thus, while the communal dinner is perceived as a great help, especially for families with children living a busy everyday life, it is equally important to balance how much time is spent in the community. As a 41-year-old female resident and the mother of two puts it:

> I actually think you are able to find that balance yourself. Yeah, I actually think so – there is no one looking to accuse you when you, for the third time in the same week, pick up food at the communal dinner and bring it home to eat. I kind of like that, in that way there is no norm or standard or so that one has to live up to, I believe. Nevertheless, we have had to invent this concept called family time. In the beginning, it was very much because the kids were outside a lot and felt like playing and all that, and then we were like 'now we also need some family time'. But now I sometimes experience, when someone knocks our door – and that happens a lot at our place – then my children sometimes say [she lowers her voice], 'Mom, won't you tell them that we are having family time', like a kind of legitimate excuse for not having to go out and play.

The concept of family time illustrates *collective pragmatism*: time with one's family is sometimes prioritised in favour of collective engagement at, for example, the communal dinner. Whereas residents express that they had a strong principle and ideal of living 'in community' before they moved into the eco-communities – an ideal that without exception exceeds the ideal of environmentally sustainable practices – they eventually compromise and balance these ideals as they experience the consequences of too much time spent in either collective or private spheres. In addition, residents experience that their collective engagement varies a lot depending on their life situations.

As expressed in statements made by residents, the question of prioritising family time occasionally affects sustainable behaviour since some practices, including cooking local produce, are more difficult to conduct individually in private homes than they are in the community. As the resident who discusses the concept of family time says, she does not experience the structures of the eco-community as being some kind of norm that one *must* live up to in one's private home. One could say that this is an unwritten yet important guideline of living in an eco-community.

The narratives about non-interference between community and private life (family time) and the attitudes towards the eating of meat illustrate a mutual understanding among the residents that living in an eco-community requires a (collective pragmatic) *balance* between engagement in the community and life at home. Despite this mutual understanding of accepting the need for balance between family time and collective engagement, some residents say that they do not find the participation in community activities 'strong enough'. Hence, there are multifaceted opinions among residents about what the right balance is and should be. Nevertheless, the fact that the general (collective pragmatic) value of not having a standard or norm for, say, how often people should participate in the communal dinner is of greater importance than the individual opinions of certain residents. Besides, the individual opinions seem to vary depending on which of the areas of collective engagement are at issue. Whereas some residents value that everyone participates in communal meetings, others find it more important to have a strong engagement in social gatherings.

Hence, the balance between community life and private life is of great importance in eco-communities. However, this premise does not imply that residents of eco-communities can chose not to participate in the community life at all and benefit from the structural advantages without being engaged themselves. On the contrary, the very liveability of eco-communities relies on residents' engagement in communal tasks and everyday social aspects. As residents often stress, 'You would not move to an eco-community if you were not interested in being close to your neighbours'.

What seems to be at stake for the residents is not so much the demand to take part in the community life, but rather that it is acceptable to choose not to participate without anyone challenging that decision. Although 'no one interferes with how often we participate in communal dinners', as one resident said, she also emphasised that her children needed a legitimate excuse for not participating in community life at all times. This illustrates how everyday life in an eco-community can never be fully separated between what is community and what is private. There will always be friends from neighbouring houses knocking on your door even though your kids do not feel like playing. You will always have to listen at communal meetings to other residents' opinions on topics that you do not find relevant, and you might even pay for meat that you do not eat. Residents might not always agree with one another on both structural questions and on social questions of participation in the eco-community. Nevertheless, choosing one's battles, compromising one's own convictions from time to time, being tolerant, and having faith in one's co-residents is the unwritten, yet crucial premise for a liveable eco-community and the maintenance of social cohesion.

Conclusion

'It has to be about people before it is about ideals', a resident concluded after a workshop where the physical and social structures of her eco-community and the question of how to maintain a liveable eco-community had been discussed by participants from a number of Danish collective sustainable movements. This resident's statement underlines the argument of this chapter: when choosing a lifestyle that enables environmentally sustainable behaviour, the consideration of the *people* in the (eco-)community must come before a focus on (collective as well as individual) ideals. In other words, when the goal is to make a sustainable transition, it is neither adequate to create only the ideals and principles for this transition, nor is it wise or even realistic to put people together believing that collective engagement alone will fulfil the principles of the community or organisation.

Social scientists have argued for a long time that so-called 'participatory community-based processes' are crucial to our understanding of how to motivate individuals into behaving more pro-environmentally (Dobson 2007, 279; Jackson 2005, 133). In this chapter, I have contributed to this notion by showing how residents of Danish eco-communities practise sustainable behaviour not through individual efforts but primarily through an organised and carefully considered system of both physical and social structures that enable the residents to develop sustainable behaviours that would often not be possible had they lived in private housing.

Sustainable behaviour is practised as a matter of course by individual residents of eco-communities, who are enabled by physical and social structures such as vegetable gardens, communal dinners, and shared responsibilities such as feeding animals. For many residents, moving to an eco-community meant that they became vegetarians, started buying fewer products, and wasted less food than when they lived in their individual housing. Living in an eco-community made these behavioural changes easier for them since they were now living within a structure that encouraged them to make more sustainable choices.

Nevertheless, depending on the community in question and the processes, experiences, and historicity it has been through, there are certain (pragmatic) conditions that determine to what degree participatory community-based processes are the solution to the promotion of environmentally sustainable behaviour. Turning to community-based approaches as an alternative to individually focused initiatives should not be an exclusive solution. Rather, the (pragmatic) balance between sustainable ideals, collective engagement, and personal considerations are crucial to making community initiatives liveable.

The sustainable structures and collective engagement of residents in the eco-communities in question are not fixed, nor do residents claim that they are perfect as they are. Instead, these communities are, and will continue to be, the result of many struggles and compromises between sustainable ideals and everyday life realities among their residents. Nevertheless, the consequences of giving up totally on these ideals is that the community loses its raison d'être. Because the consequences of not balancing collective ideals and individual freedom generate disagreements and frustrations that threaten both the community and its sustainability ideals, a compromising and more voluntary approach is the only realistic, pragmatic, and viable solution to making these eco-communities liveable. For some residents, compromising sustainable behaviour and/or engagement in the community becomes too demanding, either because it is too idealistic or because it is not idealistic enough. As a result, they move into other eco-communities or collectives that are more in line with their ideals or choose to move into individual housing where they can live as they wish, either more sustainably or less sustainably, without having to make compromises with others.

This chapter focuses on a small selection of Danish eco-communities over a specific time period. Hence, the examples provided about sustainable practices as well as structural and collective pragmatic approaches might look different had another selection for fieldwork and data collection been made. Nevertheless, many eco-communities inevitably experience similar pragmatic processes of handling problems and compromising their ideals. Eco-communities are, and always will be, ever-changing and in a state of becoming.

Notes

1 LØS: Landsforeningen for Økosamfund (The Danish Association of Eco-Communities) (see LØS 2019).
2 The majority of Danish eco-communities are not specifically called 'eco-communities'. In addition, one of my field sites is not even part of the Danish association for eco-communities (LØS) and would not necessarily even know what LØS stands for. After visiting my three field sites and other eco-communities around Denmark, I have come to realise the importance of the local and site-specificity of their names, e.g. the name of the farm building attached to the community, the name of the nearby village, or the name that was made up for the place in question. 'Eco-community' is, by contrast, rarely the utilised name: it is used to distinguish an overall category. The terms 'eco-village', 'intentional community', 'living community', or 'co-housing communities' are denominators for different kinds of sustainable settlements (Marckmann *et al.* 2012; Lockyer and Veteto 2013; Litfin 2014). According to the chairperson of LØS, the common and most precise title in the Danish context would be *eco-community* (in Danish *økosamfund*). To simplify matters I use the term eco-community here when referring to the field sites.
3 GEN: Global Ecovillage Network (see GEN 2018).
4 See Sarah Pink (2012) for reflections on living in *and* studying the everyday life of informants.
5 In Danish, 'living in *fællesskab*'. *Fællesskab* can be translated as 'community', 'social fellowship', and 'relatedness', each translation having a different weight (Bruun 2011, 62–63).
6 The term 'pro-environmental behaviour' is here to be understood as comparable to (environmentally) 'sustainable behaviour' that is used in the descriptions of the practices of the residents in the eco-communities.
7 The literature on eco-communities and sustainability discusses to what extent residents in eco-communities behave in a more environmental friendly manner compared with people living in individual housing. Among other things, the literature discusses whether the type of housing construction in eco-communities is always sustainable, and the fact that many eco-communities are often situated in the countryside and therefore generate a lot of car traffic. Nevertheless, some scholars argue that the social organisation in eco-communities upholds a certain degree of sustainable behaviour that strengthens the residents' motivation to sustain the fundamental principles of eco-community living (Marckmann *et al.* 2012, 428).

References

Berthoû, S.K.G. (2013) The everyday challenges of pro-environmental practices. *The Journal of Transdisciplinary Environmental Studies*, 12(1), 53–68.
Bisgaard, L. (2018) Private conversation. 22 April 2018.
Brinkmann, S. (2006) *John Dewey. En introduktion*. København: Hans Reitzels Forlag.
Brundtland, G.H. and World Commission on Environmental Development (1987) *Our Common Future*. New York: Oxford University Press.

Bruun, M.H. (2011) Egalitarianism and community in Danish housing cooperatives. Proper forms of sharing and being together. *Social Analysis*, 55(2), 62–83.

Dobson, A. (2007) Environmental citizenship: towards sustainable development. *Sustainable Development*, 15, 276–285.

Gausset, Q. (2010) Constructing the Kwanja of Adamawa (Cameroon). *Fractal Anthropolog*. Berlin: LIT Verlag.

GEN (2018) Global Ecovillage Network. Available at: www.ecovillage.org/ (accessed 31 January 2019).

Goldman, L. (2012) Dewey's pragmatism from an anthropological point of view. *Transactions of the Charles S. Pierce Society*. Indiana University Press, 48(1), 1–30.

Hardin, G. (1968) The tragedy of the commons. *Science*, 162(3859), 1243–1248.

Hopkins, R. (2008) *The Transition Handbook. From Oil Dependency to Local Resilience*. UK: Green Books.

IPCC (2018) Global Warming of 1.5°C. Special report. Available at: www.ipcc.ch/sr15/ (accessed 31 January 2019).

Jackson, T. (2005) *Motivating Sustainable Consumption*. London: Sustainable Development Research Network.

James, W. (1981 [1907]) *Pragmatism*, ed. B. Kuklick. Indianapolis and Cambridge: Hackett Publishing Company.

Litfin, K.T. (2014) *Eco-Villages. Lessons for Sustainable Community*. Cambridge: Polity Press.

Lockyer, J. and Veteto J. R. (2013) *Environmental Anthropology Engaging Ecotopia, Bioregionalism, Permaculture, and Ecovillages*. Oxford, UK; New York: Berghahn Books.

LØS (2016) En bæredygtig omstilling af Danmark. Strategi 2016–2019. Available at: http://okosamfund.dk/wp-content/uploads/En-b%C3%A6redygtig-omstilling-af-Danmark-L%C3%98S-strategi-2016.pdf (accessed 11 November 2018).

LØS (2018) Private conversation with boardmember Niels Aagaard. 3 April 2018.

LØS (2019) Landsforeningen for Økosamfund. Available at: www.okosamfund.dk (accessed 31 January 2019).

Marckmann, B. (2009) Hverdagslivets Kritik – Økosamfund i Danmark. PhD thesis, Department of Sociology, University of Copenhagen, Denmark.

Marckmann, B., Gram-Hanssen, K., and Christensen, T.H. (2012) Sustainable living and co-housing: evidence from a case study of eco-villages. *Built Environment*, 38(3), 413–429.

Miller, F. (2018) *Ecovillages around the World: 20 Regenerative Designs for Sustainable Communities*. Rochester, VT: Findhorn Press.

Mogensen, H.O. and Whyte, S.R. (2004) Antropologi og Medicin i Dialog. *Tidsskrift for forskning i sygdom og samfund*, 1, 39–58.

Pink, S. (2012) Researching practices, places and representations methodologies and methods. *Situating Everyday Life: Practices and Places*. London: SAGE.

Reinholdt, L. (1997) *Bosætningseksperimenter*. Denmark: Svanholm Forlag.

Shove, E. and Spurling, N. (2014) *Sustainable Practices: Social Theory and Climate Change*. New York: Routledge.

Shove, E., Pantzar M., and Watson, M. (2012) *The Dynamics of Social Practice. Everyday Life and How it Changes*. London: SAGE.

Udlændinge-, Integrations- og Boligministeriet [The Ministry of Immigration, Integration and Housing] (2016) 'Fremtidens bofællesskaber i funktionstømte bygninger i storbyen, provinsbyen og på landet'. Denmark, Dansk Bygningsarv.

United Nations (2019) About the Sustainable Development Goals. Available at: www.

un.org/sustainabledevelopment/sustainable-development-goals/ (accessed 21 February 2019).

Whyte, S.R. (1997) Questioning misfortune: the pragmatics of uncertainty in eastern Uganda. *Cambridge Studies in Medical Anthropology 4*. Cambridge: Cambridge University Press.

4 Stronger together

How Danish environmental communities influence behavioural and societal changes

Quentin Gausset

Background

Denmark has one of the highest environmental footprints on earth (WWF 2014; Strobel *et al.* 2016), and there is a broad consensus in the country that a sustainable transition cannot be achieved without large-scale changes in behaviour and consumption patterns. But, to date, policies designed towards changing behaviour target individuals almost exclusively, mainly through information campaigns and economic incentives such as green taxes and subsidies (Darnton 2004; Darnton *et al.* 2006; DEFRA 2005, 2008; Jackson 2005; Lucas *et al.* 2008). These policies rely on a range of psychological or economic theories that explain behaviour as the result of individual choices made to maximise benefits and minimise costs. They include the Rational Choice Theory (North 1990; Abel 1991; Allingham 1999), the Expectancy Value Model (Fishbein 1967, 1968; Fishbein and Ajzen 1972, 1974), the Theory of Reasoned Action (Ajzen and Fishbein 1980), the Theory of Planned Behaviour (Ajzen 1985, 1991), Behaviourist Theory (Skinner 1938; Ferster and Skinner 1957), and Social Learning Theory (Bandura 1963, 1977), to name just some of the most influential. All these theories and the current environmental policies inspired by them understand behaviour primarily as a question of individual choices that are based on costs and benefits (or punishment and rewards) associated with different choices and on the information that individuals have about these outcomes.

I argue in this chapter that behavioural and societal change should also be understood as the result of social and political interaction, i.e. as something that is primarily social and political in nature, politics being here understood as the realm of allocation of resources/duties and negotiation of conflicts. Social scientific approaches locating change in social interaction and in collective action must therefore complement psychological and economic approaches locating change in individual choices.

The data on which this analysis is based come from (a) interviews at and field visits to four eco-villages or low-energy housing communities (the cooperative society in Hjørtshøj, the self-sufficient village in Hundstrup, Munksøgård, and Svalin in Roskilde), two environmental fairs ('Rethink Activism' in

Århus and 'The Common Good' in Vejle); (b) interviews and participant observation in one collective permacultural garden in Køge; (c) three focus group discussions at the association 'Transition Now'; (d) a course on sustainability in Brenderup folk high school; and (e) interviews of various key actors in the food cooperatives in Copenhagen, Køge, and in other environmental associations, such as the Green House in Køge, the food dumpsters' mobile kitchen in Århus ('Skraldecaféen), and the head of CSR for COOP (a Danish cooperative chain of supermarkets). The general argument developed here has been inspired by a number of researcher seminars organised within the Compass research project (www.compass.ku.dk).

In the aftermath of COP15 in 2009 and the failure of state actors to tackle climate change issues, citizens are increasingly taking matters in their own hands (Hoff and Gausset 2016) and increasingly organise themselves in environmental communities. The communities discussed here are all constituted by people who would like to live more sustainably by changing their way of life and their consumption habits. These people may have different starting points, however. Members join food cooperatives to access organic, local, and seasonal food. Residents of eco-villages try to reduce their environmental footprint by relying more on renewable energies, by reducing consumption and increasing sharing, and by building passive, biodegradable or off-grid houses. In the Transition Now association, members reflect upon personal and societal barriers to a sustainable transition so as to overcome them. Yet, despite these different goals and *modus operandi*, these environmental communities differ from other more classic environmental organisations in that they focus on changing their own lifestyles at the grassroots level rather than on changing other people's behaviour or society at large through policy change. Their focus is more inward than outward. I have therefore chosen to refer to them as environmental *communities* (*fællesskab* in Danish) rather than as environmental associations, organisations, or movements. This concept of community has the advantage of implying close and regular interactions among its members as well as a strong convergence of norms and values among members.

Environmental communities as physical and socio-political infrastructures

A community is usually characterised by a shared environment and infrastructure. In the cases I have studied, one of the primary motivations to create or join an environmental community is a general frustration regarding mainstream society and a desire to live according to an alternative and better model and in a different type of environment, society and/or infrastructure. People join an environmental community because they wish to access the services it provides: organic food in food cooperatives or in urban gardens, the safe and healthy environment in eco-villages that stress solidarity among its residents, and environmental information and networks in 'generalist' environmental communities. These services are provided by specific physical and social infrastructures.

The primary drive for creating organic food cooperatives in Denmark has been the difficulty accessing organic food in supermarkets and the necessity of joining forces to organise a parallel system of organic food supply connecting producers in the countryside to urban consumers. In the words of one of the founders of a food cooperative in a small town of Zeeland,

> Fifteen years ago, when we started the cooperative, there was no organic food in supermarkets. Things were a bit better for people living in Copenhagen, but outside the capital, it was really difficult. The only way to access a variety of organic farm products was through our cooperative. Today, things have a changed a lot, but back then, we were pioneers.

Likewise, the primary motivation for creating eco-villages has been the wish to live more self-sufficiently and more sustainably. For people who want to experiment with sustainable housing built with local and bio-degradable materials (hay, wood, mud, mussel shells), off-grid housing disconnected from a sewage system (developing the first ecologically-based purifying systems), self-sufficient housing relying on renewable energies (solar, wind, wood), and a self-sufficient lifestyle by growing their own food, it is easier to join forces in an eco-village with a common sustainable infrastructure than to create everything alone. It is much easier to engage collectively with the municipality and have them change local plans so as to accommodate more sustainable buildings and infrastructure. For example, building codes could allow some agriculture in residential areas or some buildings in agricultural zones (whereas traditional residential and agricultural zones tend to be clearly demarcated and exclusive of one another), could accept buildings made of straw and mud that defy conventional building regulations, or allow off-grid buildings disconnected from the sewage system and using natural purification systems. Living more sustainably often challenges municipal regulations, and collective organisations have more leverage power than individuals to deal with this problem.

Thus, working in groups is often a prerequisite for accessing goods and services that are out of practical or financial reach for individuals because it is one of the only ways through which Danish citizens can (or could, at some point) access local organic products or live more sustainably. It is by pooling efforts and resources together that groups of people can design new infrastructures and new communities characterised by a more sustainable way of life.

Elisabeth Shove and her colleagues (Shove 2010; Shove *et al.* 2012) argue that daily practices are not so much determined by individual choices as they are influenced by the physical infrastructure that frames the daily lives of people – an idea that one also finds at a lower scale of analysis behind the Nudge theory (Thaler and Sustain 2008). These authors draw attention to the necessity to change the infrastructure in order to change daily practices. But while these scholars see infrastructural change as something driven from above and piloted by governments or enlightened authorities, members of food cooperatives and eco-villages do not wait for governments to change infrastructure. They take the matter in their own hands and design new physical and economic infrastructures such as sustainable energy supply systems, organic food distribution chains, or creating new possibilities for reusing and recycling old goods, among other things.

It is important to recognise that infrastructures are not just physical and economical, they are also social and political. All collective associations develop institutions to manage common activities and common goods. I refer to these institutions as a socio-political infrastructure because they represent a basic structure that organises social interactions in the community. First, they typically include some kind of micro-government in the form of an elected board, general assemblies, and legal statutes defining the overall goals of the community and procedures for making decisions. Agreeing on how to manage common activities and resources is not an easy task; it is characterised by various kinds of conflicts. As a consequence, environmental communities often experiment with different forms of direct democracy and increasingly also with sociocracy in order to run their communities and manage conflicts.

Second, environmental communities rely heavily on voluntary work. They can only function if members bear the collective costs of running the community and give part of their free time to make it work. Danish food cooperatives, for example, do not just sell organic food packs but also require that each member gives three hours of volunteer work per month to distribute organic food in order to keep prices low. Most members use these hours to run the weekly delivery of food, while others get involved in food transport or in the administration of the cooperative. In eco-villages, too, one usually finds several working groups that are responsible for managing the common activities and resources of the community (meals, food production, laundry, buildings, festivals, etc.). Defining the rights and duties of members, including the membership fee and the amount of labour and engagement that each and every one must provide to run the community, is an important aspect of the socio-political infrastructure.

Infrastructures curtail behaviour. Physical infrastructures not only provide access to greener services (for example, a local system of district heating reduces dramatically the carbon footprint associated with heating houses), they also influence behaviour and can push it in a more sustainable direction. For example, ecological sewage systems using ecological processes for water processing and nutrient recycling limit dramatically the amount of chemicals that can be thrown in the sewage. A large list of products harmful to the environment cannot be used because they cannot be naturally filtered. Interestingly, when one of the eco-villages that I visited during my study temporarily connected its sewage system to the conventional municipal 'grey' system while it was renovating its ecological sewage system, some members rejoiced that they could now dye their hair, which was a forbidden practice before because these kinds of beauty products cannot be filtered naturally. This is just one example of how physical infrastructures can be much more influential than personal consciousness in curbing environmental behaviour.

Socio-political infrastructures also play an important role in influencing behaviour because members of eco-communities must abide by the overall goals defined in community statutes and must accept the rules and duties described either formally in by-laws or informally in the daily activities of the community. Communal meals, for example, are an important part of the community life in eco-villages, and the rules defining the type of food consumed can be more or less environmentally friendly. In some villages, meals are vegan, while elsewhere they are simply vegetarian, and in yet others meat can be consumed as long as it is produced in the eco-village (poultry, sheep, or cows). Even though individual families are usually free to consume whatever they wish in their private homes, food in collective meals is a very sensitive issue and the object of much debate. For some members, disagreement on food ethics is a sufficient grounds to leave a community and join a new one that is closer to their ideals.

As we can see, people wanting to become members of an association must abide by existing structures and align their expectations and behaviour with the rules that have been designed by the first members and adapted further over time through different conflicts and adjustments. Just as physical infrastructures organise daily practices and consumption patterns, socio-political infrastructures define values and norms that organise how members behave and interact. Being a member of an environmental community thus reduces the freedom of each member to behave as s/he wishes: it is the price that one needs to be willing to pay to benefit from the advantages linked to group membership. But this is also one of the mechanisms that influences peoples' behaviour and makes it more sustainable.

Understanding social norms as part of a social infrastructure has the advantage of drawing attention to norm design and malleability. Norms and values are not just created 'from above' and imposed through socialisation, as implied by classical sociologists such as Emile Durkheim (1950 [1895]), George Herbert Mead (1934) or Talcott Parsons (1968 [1937]). They are also created and recreated through social interaction. This is evident when new environmental communities are created: the first members spend many, many hours defining the values and the norms that will constitute the community's foundation, and the first years of existence are typically characterised by deep debates, conflicts, and identity crises that lead to continuous adjustments of the original values and norms.

Viewing social norms as part of a socio-political infrastructure also has the advantage of explaining why behaviour change can appear undramatic or can even go unnoticed. This is a striking feature of many members of environmental communities, who tend not to see themselves as especially environmentally minded or who often profess a strong feeling of guilt for not being more environmentally minded than they perceive themselves to be. 'We are not doing anything special, we just live here' is a common answer to questions relating to their environmental behaviour (see more on this in Høite Hansen's chapter in this volume). Even though members of environmental communities do have a more sustainable lifestyle than non-members, they often do not stress this because their greener practices essentially derive from physical and socio-political community infrastructures to which they adapt or conform rather than from personal choices and efforts. The most important personal choice lies in joining an environmental community in the first place. Once this choice is made, the member then follows the choices that have been made by others without reflecting much upon them.

This point confirms Elisabeth Shove's observation that practices are seldom reflected upon and derive almost 'naturally' from infrastructure. Where my analysis differs from Shove is that the infrastructure in environmental communities is found at the micro-level. Moreover, it is not just physical, it is also socio-political and is developed from below rather than from above.

Environmental communities as collective identities

Another major motivation to join environmental communities is to avoid feeling lonely and to be together with people who share the same norms and values. It is much easier to display a sustainable behaviour when one is not the only one doing so. Talking to like-minded souls reinforces individual choices. A recurrent theme among my informants is the frustration and loneliness that they felt and resented feeling before joining an environmental community and the enthusiasm they felt after joining a community of people that shares the same environmental ideals. Joining such a community is experienced as a great relief and an emancipation for many people as it reinforces their belief that their personal and societal choices are legitimate because other people shared them. Humans are moral beings: they want to be good and do the right things. But when doing the right thing implies going against mainstream lifestyles, that is, against mainstream eating, clothing, transport, housing and/or consumption habits, it is much more difficult to do it alone and be considered a lunatic or marginal by mainstream society. For many of my informants, it is far easier to do the right thing as part of a group of like-minded people and be considered part of a visionary and frontrunner movement.

One of the consequences of joining an environmental community is that people become more daring in exploring new dimensions of sustainability. By entering a green community, they engage on a journey that will slowly make them progress on a path that starts with the official goals of the association (organic food, self-sufficient lifestyle) but then leads them to adopt ever more environmentally friendly practices. These more 'advanced' sustainable behaviours can include eating more seasonal, local, vegetarian, vegan food, wearing organic clothing, using renewable energies, and re-using, sharing, and reducing a wide range of things, such as waste, packaging, travel, debt, credit, or consumption. This journey is guided by the different community members who share the same values and ideals and who are all struggling with the same challenges along the road. For example, several informants from the Copenhagen Organic Food Cooperative remember their embarrassment when they were asked during one of their first encounters with older members what their favourite seasonal fruit or vegetable was and giving the name of a vegetable or fruit that was out of season. New members of the food cooperative thus quickly develop an interest in more than simply organic food, including aspects such as seasonality, locality, transport, fair-trade, and so on. Is it ethical to buy avocados from Israel or to buy vegetables packaged by prisoners? These discussions are important and generate hot debates not only by the board of the food cooperative in charge of ordering food but also among the cooperative's general membership. If these issues are not addressed properly, such conflicts can lead to a haemorrhage of members or a schism within the community.

Such dynamics are classically understood by referring to concepts such as conformism and peer pressure, i.e. in terms of rewards for those who conform to social norms and sanctions for those who deviate from social norms (Festinger 1954, 1957). My research shows that although rewards and sanctions are present, they do not dominate people's narratives – there is much more at play. When they talk about peer pressure, several informants admit that the sanctions against them are more imagined than real (some talk about 'silent judgements'), and that they often behave according to how they *think* their peers would judge them without having ever experienced any actual judgement. One informant explicitly downplays the importance of environmental peer pressure by saying that 'people don't like me because I am environmentally conscious; they like me because I am a kind person'. Too much focus on conformism in a community can also push members away, which is seen as a problem and is therefore avoided. So even though it is true that there is some pressure to conform to the community's ideals, this pressure is primarily limited to collective activities, and there is a large tolerance for deviance. For example, in eco-villages that organise collective vegetarian meals, meat is also served occasionally for meat-eating families. Even when collective meals are fully vegetarian, some children might wish to eat hamburgers or sausages at home once in a while. Although this goes against the ideal of the community, not much fuss is made about it, and a family can then choose to retreat from a collective meal to stay at home and eat a meal that contains meat (see Høite Hansen's chapter in this volume).

Interestingly, a recurring theme among informants actively involved in environmental associations is that they very consciously try to avoid 'predicating', that is, telling people what they should do. They are just open about their own life choices and make them known to their relatives and friends, which means that they don't feel the need to explicitly disapprove of choices that go against theirs. Informants report that it is often people who hold mainstream behaviour who feel the need to apologise for not living up to higher ethical standards, saying, for example, 'Sorry, I know that this is unethical and that you disapprove of this, but I do it anyway because I am used to doing it'. Thus, without expressing any disapproval or any comment on other people's behaviour, members of environmental communities who live up to their sustainable ideals do influence those with whom they interact by the sheer fact that their interlocutors feel compelled to compare their behaviours, and also because their interlocutors often share the same ethical ideals even though they do not live up to them to the same extent.

In the words of one female informant in her late twenties, interviewed in a focus group discussion (FGD):

> Several of my boyfriend's friends believed that I was vegan, although I am flexitarian. Because I was so green, they believed that I must be knitting all my clothes by myself and that I refuse to use any transport means except bicycle, which is not true at all. But slowly by slowly, in all the time I have been with my boyfriend, I have sensed that this green thing, which I came with, has slowly penetrated them. And then suddenly they come and tell me joyfully that 'I have now ordered my first parcel from Årstiderne [a private company delivering packs of organic food at home], and I have come to like vegetarian food'. Me, I don't say a word, but they know very well where I stand. If one is open and clear about this [one's green values], then one does not need to moralise others to influence them.

Another female informant in her late twenties continued along the same line in the same FGD:

> Before I was part of it [Transition Now, an association working for a sustainable transition of society], when I expressed these [green] opinions, I was classified as an extremist and radical person. But I can actually feel that being a member of Transition Now has given me a mandate to say that I am not alone with this. There is actually a big network of people who come from very different backgrounds but who all think the same. And then, [hearing that I was a member of the association], worried people began to ask me about different things: 'Could you explain me this or that?' They could see that I was not at all an extremist who was predicating, but that I was a member of an association. People were much more willing to listen to me than when I was alone.

And, to conclude, the first informant also said:

> Yes, one can easily influence without moralising. And I also think that it is easier if you know that you are a part of a community. We [in Transition Now] all agree that we are on to something really important. And then I don't need to lift my index finger [to predicate] when I sit in a dinner with my in-laws. They know very well where I stand or they can ask about it. They have done this a lot. I think that it is easier to communicate one's message in this way.

To my surprise, I found less peer pressure in environmental associations than what I had expected from reading the scientific literature. Classic behaviourist theory argues that people behave according to rewards and punishment associated with specific behaviours. It focuses on conformism to pre-existing norms and explains this mainly through peer pressure or through institutionalised systems of rewards and punishment. However, as we have seen above, peer pressure and social conformism are only a small part of group dynamics, a part that is explicitly downplayed by several actors (see also Van den Pol 2017). Understanding behaviour as a response to external stimuli misses the process through which new norms are created without sanctions by simply stating strong and steady opinions and choices, by reflecting upon other opinions and choices, and by eventually changing in the process. Thus, attitude and behaviour change happens without much conflict or reward/punishment as a result of the sheer interaction and alignment of people holding different values and comparing each other's opinions. In the words of one informant reflecting on this process: 'It is really funny to see how one can change things. It is crazy how little is needed to get things to change'. It is thus possible to influence others without criticising them for their 'bad' behaviour just by stating clearly and openly where one stands, which makes the disapproval and related peer pressure implicit rather than explicit – if it exists at all. Being a member of an association makes this kind of statement much easier to profess and triggers larger scale changes.

An important aspect of this analysis is the freedom of association. While classical social psychology and economics base their theories on the prerequisite that people are trapped in society and incapable of escaping from the established social game (i.e. that people are forced to choose between being rewarded or being punished), real life in Western societies is much more complex. People can escape the rules of a social game by creating (or joining) a sub-community in which they can behave according to rules other than those found in mainstream society. This accounts in part for the limited relevance of rewards and punishments in explaining people's environmental behaviour.

Another remarkable aspect of environmental communities is that their members, because they are more concerned with changing their own behaviour before changing that of others, seem to display more tolerance towards other people's behaviour as compared with people belonging to more activist organisations. In other words, research data suggest that when political activism and the wish to influence other people's lives or society gains in importance, tolerance shrinks. For example, vegans are both noticeably more activist than vegetarians and at the same time also much more 'predicating' and 'moralising', or less pragmatic and less tolerant than vegetarians when interacting with people having different food preferences.

Environmental communities as educational institutions and incubation labs

A third kind of motivation to join an environmental community is the wish to learn something, meet new people, and create new experiences that support one's interest in sustainability. We have already seen above how people who join an environmental community for a specific purpose end up becoming interested in a larger range of issues than their original motivation. Members of a community engage on a learning path in which they try something new and make personal progress. A corollary of this point is that once people have created new experiences and have acquired new knowledge and networks or once their life and family situation make it more difficult for them to learn something new, they are more likely to leave the community. When the main motivation to join an environmental community is to meet new, interesting people and to acquire new experiences before trying something else, then members' commitment is not a lifetime engagement but is instead for a limited period of time. As one informant who has been inactive for a year and who is on her way out of Transition Now explains:

> I think of it [Transition Now] more as an incubator. If I think of my own history, I feel like I entered Transition Now because I wanted in some way to work or do something with transition and this was a place to start. And then I met people and I did things and got inspired and for sure that pushed me in the direction that I am in now.

The 'direction in which she is in now' is that this informant has dropped out of university and has begun training to become a carpenter. This situation is far from unique among young members of environmental associations: it is not uncommon to meet young members whose engagement in an environmental community has led them to drop out of higher academic education to engage in a more practical or manual education related to agriculture, gardening, or building, among other things. When asked if she would have chosen that path had she not been involved in 'Transition Now', she answered:

> Probably not. I think that it [Transition Now] worked like an incubator. You get inspired and you go different ways and try different things and you get the courage and everything. And then you figure out: 'OK, now I am ready to fly'.

Members who leave an environmental community do not forget about the knowledge, experiences, and norms and values acquired during the period of their membership. On the contrary, when leaving the community, they often share them in their interactions with other people. I have recorded several instances of people who left an environmental community with the specific purpose to create a new one or start an environmental or sustainable business (a restaurant serving organic and local food, an enterprise producing organic juices or building low emission houses, a shop selling organic or bio-dynamic products, etc.) or engage in environmental politics (see more on this below). This means that people who move out end up disseminating the community's goals in other spheres of society, which multiplies by many times the environmental impact of the community. A striking result of our research is the genealogical ties connecting environmental communities: most of them can be said to have emanated from other communities since they have often been created by members who left older communities – the first ones dating back to the 1970s. Environmental communities are thus linked through rhizomatic chains, and they spread their ideas in part when their members leave them to create new communities. This kind of growth is made possible because environmental communities are places where knowledge, experiences, and networks are created and transferred to members.

Since environmental communities can be seen as some form of educational institutions or incubation labs influencing behaviour through the production and transfer of knowledge, experience, and network, it is not so surprising to see that this type of knowledge and experience is being increasingly integrated into alternative educational institutions. For example, environmental communities are disseminating their knowledge and skills in the existing network of Danish folk high schools, or *'højskoler'*, non-formal residential schools that are very popular in Denmark and offer non-degree learning opportunities for adults. The folk high schools that provide courses on sustainability recruit teachers who come from environmental communities and who transfer their experience to students/participants. Environmental communities host a large number of people who have become true experts in very specialised sustainability niches, such as permaculture, ecological sewage, renewable energy, off-grid building, etc. This expertise is increasingly recognised not just by folk high schools, but also by local governments, businesses, and even banks (some members of eco-communities have been asked to sit in the board of banks to monitor their environmental and corporate social responsibility records).

One might wonder whether communities attract people who are already knowledgeable before they join or whether new members become more knowledgeable after they join an eco-community. After all, most members of eco-communities are so-called 'resource-strong' citizens, i.e. people with higher education and/or a strong engagement in civil society. But research results show that the knowledge and expertise needed to run an eco-community pushes people who might already have a lot of expertise into new directions, forcing them to specialise in new fields of relevance for their community, which broadens and advances their knowledge. As one informant puts it: 'No matter how much people know before they join an eco-village, they will *always* make tremendous progress and learn much more about sustainability'.

Knowledge production and transfer are explicit goals for most environmental communities, and they are important aspects through which their collective action makes an impact on society.

Environmental communities as political lobbies

A fourth motivation to join an environmental community is to join forces in order to have more impact on society. People interested in sustainability are generally not just selfishly interested in improving their own life but also have a more general and altruistic desire to improve society and the planet (Gausset *et al.* 2016). Creating or joining an environmental community can be a step on that path.

It is rather difficult to measure the impact that environmental communities have on society at large in Denmark. Many environmental policies have been influenced by discussions in informal settings, informal networking, lobbying behind closed doors, participation in the media, and debates, but the causal relations linking these activities to societal impact are difficult to document with classical social scientific methods. Nevertheless, the role of environmental associations in triggering the recent shift towards much more urgent and more prioritised environmental policies (as compared to just one decade ago) should not be underestimated.

Few people took eco-villages seriously when they experimented with wind turbines in the 1980s. But they were front-runners, and this technology has now become a mainstream source of renewable energy that has become a major export industry, employing more than 32,000 people in Denmark (Windpower 2016). Few people took eco-villages seriously when they started growing organic food back in the 1980s, but organic food can now be bought in all kinds of supermarkets. One can discuss to what extent wind energy or organic food really gained ground in Denmark because of environmental communities or whether these new trends were part of a global movement that would have hit Denmark anyway, with or without eco-villages and food cooperatives. But one can safely assert that environmental communities have been environmental front-runners and trend-setters; there are arguments supporting the thesis that these organisations have played an instrumental and crucial role in pushing for a sustainable transition in Denmark.

Eco-villages were among the first to practise organic farming, for example, well before it became a common farming practice. Food cooperatives were also among the first to organise the supply of organic products in urban settings well before they became common goods in supermarkets. By doing so, they demonstrated that mass distribution of organic products was feasible and that there was a real demand for them, which paved the way to mainstream production and commercialisation.

Environmental communities swam against a strong current when they started. They had to work against prejudices and against existing regulations and they had to find ways to overcome all sorts of practical, legal, and political barriers. But the principles designed by the first environmental associations producing and distributing organic food ended up having a direct influence on national policies. For example, the eco-community of Svanholm was among those who took the initiative to create the first national association of organic agriculture (*Landsforening for økologisk landbrug*, created in 1981). Svanholm also played an instrumental role in defining the first rules characterising organic farming and designed the first logo certifying organic food (Hansen and Andersen 2015). These rules and regulations then ended up constituting the basis for the Danish national legislation on organic farming in 1987 and the State-based certification of organic food in 1990.

Even today, after the introduction of the European Union's (EU) organic farming certificate and logo, Denmark continues to use the standards defined by its national association of organic agriculture in the early 1980s, as these standards are higher than those defined by the EU (Økologi 2015). The development of such standards was never easy: according to Paul Holmbeck, the director of the new land-based association for ecology organic products, it was the result of three things: 'fight, fight, and fight. Again and again' (Hansen and Andersen 2015, 15). Such achievements have only been made possible by citizens joining forces and creating the first eco-communities, then joining forces with other organic farmers in a land-based association, and then lobbying politicians to make room for organic products and promote them. Today, Denmark has the highest organic share of the total food market in Europe (Willer 2017, 13).

The success of environmental communities in influencing mainstream society is such that it can become a problem when their once innovative ideas and practices spread into all strata of society and when environmental communities are no longer at the forefront of environmental action. This pushes environmental communities to be always on the move towards newer and more profound forms of environmental engagement. For example, as supermarkets are now competing with each other to provide the best and cheapest organic food, food cooperatives no longer play the leading role that they used to play at a time when accessing organic food was difficult. Membership has been dwindling and food cooperatives are now discussing how they can reinvent themselves. Likewise, organic farming is no longer the panacea in eco-communities as this form of farming can have a higher carbon footprint than non-organic farming. Consequently, the focus of environmental communities has shifted away from simply organic farming to permaculture or biodynamic farming and a stronger focus on regenerating degraded soils and developing farming systems that capture more carbon than they release.

A third example is that the increased municipal focus on sustainability has led to the development of efficient waste sorting and CO_2-neutral district heating systems. As a result, several eco-villages, which used to be the front-runners in their use of sustainable heating or recycling systems, are now lagging behind what is available elsewhere in their municipality and are discussing whether they should join the municipal network or whether they should develop new cutting-edge approaches that outperform municipal infrastructures.

Three mechanisms combine to explain the influence that environmental communities have on environmental policy-making. First, front-runners and fiery souls who lead and drive environmental associations often contact local governments for practical reasons. For example, an eco-community wanting to buy a piece of communal land or conduct small-scale farming needs to talk to the municipality about the different restrictions in existing local plans and negotiate an agreement to make things work. A housing cooperative that wants to become CO_2-neutral can get into trouble with the local plan of the municipality which prevents the use of solar panels in the centre of town and must engage in a dialogue with municipal agents to find solutions. Some food cooperatives wish to organise weekly collective meals but face problems if their kitchen does not satisfy the latest requirements regarding public catering or if a municipality considers that this constitutes an unfair advantage over existing catering businesses. In the city of Århus, a community of dumpster divers, which serves free 'wasted' food in a mobile restaurant, faces expensive and bureaucratic requirements from the fire department about the endorsement of their kitchen each time they move to a different social event. They are also not allowed to serve dumped food to people, but get around this prohibition by inviting people who eat the food to prepare it together with themselves (as the legislation still allows people to eat whatever food they prepare for themselves).

As we can see in these numerous examples, environmental associations constantly challenge existing regulations. Sustainable transition might very well be on every politician's lips and prioritised at the municipal level, but the reality experienced by environmental communities is that there is a big gap between what is declared and what is practised. Each and every environmental organisation can tell stories about how local or national governmental regulations have constrained their actions and how they have had to fight against the system or to play with grey legal zones in order to make room for more flexibility. In this process, environmental communities enter into a dialogue with local authorities to find creative solutions acceptable to all. Municipal agents are often torn between enforcing the rules strictly and supporting the environmental initiatives that fit nicely with municipal environmental goals. On certain issues, municipal agents adopting one approach oppose some of their colleagues who adopt an opposite one, and an internal struggle ensues within the municipal administration. Because of all the obstacles they need to overcome, environmental communities are slowly influencing regulations to make them more amenable to environmental action, and the increased flexibility that results also benefits the general population.

A second factor explaining the political influence of environmental communities is that fiery souls who spend time and energy to create and manage environmental communities are performing some form of local politics within their environmental communities. In other words, they are de facto more politically active than other members. The line between local politics in an environmental community and local politics in a municipality is a thin one and one that is easily crossed. This point is connected to the former one: the frustration derived from interacting with authorities and experiencing them as a barrier rather than as a motor for change is a big motivation to engage in politics. As an environmental activist who was elected as municipal councillor in the last municipal elections of 2017 explained:

> When we were trying to establish a transition town, we engaged in a dialogue with the municipality but we ran into a wall. We then agreed that one of us had to engage into politics and the choice fell on me.

For someone who wants to make a difference and is prevented from doing so by 'the system', it makes sense to engage in politics and change things from the inside. Fiery souls have acquired experience in the daily management and organisation of environmental communities, they have demonstrated leadership, organisational and/or communication skills, and they have gained self-confidence in this process – all qualities that are important in politics.

Moreover, managing an environmental community means being part of a larger network of communities, i.e. knowing (and being known by) many people who share the same vision and having access to a network of people who can open many doors and confer strong advantages in the electoral process. To summarise: fiery souls are often idealists with community management experience who want to build a better world, which pushes them to engage in higher scales of politics in order to have stronger leverage and make a bigger difference.

In the political engagement of these activists, the environment is a prominent focus which cuts across party politics, which means that they can easily talk to people from different political parties who share the same interest. Even though members of environmental communities tend to lean to the left side of the political spectrum, they nevertheless are generally spread across the political spectrum and engage in competing political parties. The consequence of this political plurality is that the same environmental agenda can end up being shared by opposing political parties, which multiplies the likelihood of its being adopted. The picture emerging from this is not one that pits environmental engagement against traditional political engagement or that tries to explain the decline of traditional political involvement as being due to the disappointment of citizens in political institutions (as argued, for example, by Norris 2003 or Putnam 2000). On the contrary, it points to environmental goals as a new kind of motivation to engage in the traditional political system.

A third mechanism by which environmental communities influence politics is through the voice that they gain and the attention that they raise. Environmental organisations are gold mines for the media: they come with ideals and utopias, they claim to be the only drop of reason in an ocean of madness, they display exotic behaviours that swim against the current, they provide photogenic houses, landscapes, and lifestyles, and they deliver well-articulated and charismatic thinkers that can give hope in a time of despair and urgency. All this has a strong resonance among the media consumers, and therefore also among policy-makers. In the words of an environmental leader who has engaged in local politics:

> It is not so difficult to make a change. You just need to make noise. If the politicians feel that people want change, they will deliver it [...]. The solution is to organise in communities. If politicians see that you are organised, they have to listen.

The size of an environmental community and its coverage by the media converge to give the community visibility and political weight, which allows it to be heard more easily. The interest of local politicians in environmental communities is twofold. On the one hand, politicians are always interested in finding popular citizen support for their policies. Since environmental communities are more vocal than 'normal' citizens, their voice is stronger and it counts more. On the other hand, communities have accumulated experience and have developed networks that can be interesting for and useful to policy-makers. One example described above is the Danish state-controlled organic certification that derives from a system initiated by eco-villages. Another example is the environmental policy of the new political party 'Alternative' (the youngest and greenest Danish political party) that was written by a person who was recruited for his vision and expertise deriving from his engagement in the Transition Now community. Another example is the chart of the 10 core principles defining Copenhagen food cooperatives that has directly inspired the food policy of public restaurants in several municipalities. These examples show how local politicians and policy-makers have become interested in collaborating with different environmental communities, in using their expertise and sometimes even in co-opting them to capitalise on their network and experiences.

Most environmental communities discussed in this chapter do not see their primary goal as lobbying and influencing national and international politics (unlike Greenpeace or the World Wildlife Fund, for example) and focus more of their attention on changing their own daily life and consumption than on influencing society at large. However, they definitely have an important impact on environmental policies through the interactions they have in their daily activities with policy-makers, through their direct engagement in local politics, and through the strong collective voice deriving from their numbers and expertise. Thus, although environmental communities define themselves as apolitical, they are all deeply involved in some form of politics (see more on this in the Hoff and Islar chapter in this volume) and they impact on both society and the environment. They exert their influence through discrete individual connections and political engagements rather than through the louder collective demonstrations, activist happenings, or professional lobbying that is characteristic of the more established environmental NGOs (see, for example, Wapner 1996).

Discussion and conclusion

Environmental communities influence environmental behaviour through at least four different mechanisms: (1) through designing physical, social, and political infrastructures that influence everyday-life choices; (2) through collective dynamics characteristics of collective identities in local communities; (3) through producing and transferring knowledge and experience; and (4) through political engagement and lobbying. Environmental communities can thus be seen as social groups and identities, as new infrastructures, as educational institutions, and as political institutions. All these aspects are grounded in collective action rather than in individual choices and confer credit upon environmental communities as having a significant impact on environmental behaviour.

If is true that attitudes influence behaviour (see for example Henn and Kaiser in this volume), little is known about how attitudes are made. This chapter suggests that the roots of environmental attitudes are to be found in community and society, i.e. in the way ethics, values, and norms are negotiated through social interactions. While classical scholars studying norms and values focus on their top-down imposition through the socialisation of children, this chapter focuses on how new norms and values are created in eco-communities in a bottom-up negotiation. Seen from this perspective, environmental behaviour is not just a question of responding to stimuli such as threats of punishment or promises of rewards. It is more a question of creating and negotiating norms and re-creating and re-negotiating them constantly while interacting with others, being inspired by other people's ideas and actions, and positioning oneself in relation to them. The creation of new norms happens as a simple consequence of the need to find common ground for social interactions in a group, without any recourse to rewards or punishment (other than accepting or refusing to engage in social interactions).

This chapter argues that the classical understanding of environmental behaviour as the result of individual choices must be complemented by a new understanding of environmental behaviour as the result of collective choices negotiated through community-building, collective identity, education, and political engagement. Social and environmental change will not just come from more (top-down) information and incentives allowing individuals to make better cost-benefit calculations; it will also come from (bottom-up) collective action and community-building. Since the current environmental policies focusing on individual choices have largely failed to bring about a widespread change in behaviour, policy-makers would be well-advised to also facilitate collective action in eco-communities. Environmental communities are natural allies of environmentally minded policy-makers, and collective action is one of the keys to a more sustainable future.

References

Abel, P. (1991) *Rational Choice Theory*. Aldershot: Elgar.

Ajzen, I. (1985) From intentions to actions: a theory of planned behavior. In J. Kuhl and J. Beckman (eds), *Action-control: From Cognition to Behavior*. Heidelberg: Springer, 11–39.

Ajzen, I. (1991) The theory of planned behavior. *Organizational Behavior and Human Decision Processes*, 50, 179–211.

Ajzen, I. and Fishbein, M. (1980) *Understanding Attitudes and Predicting Social Behavior*. Englewood Cliffs, NJ: Prentice-Hall.

Allingham, M. (1999) *Rational Choice*. Basingstoke: Palgrave Macmillan.

Bandura, A. (1963) *Social Learning and Personality Development*. New York: Holt, Rinehart, and Winston.

Bandura, A. (1977) *Social Learning Theory*. Oxford: Prentice-Hall.

Darnton A. (2004) Driving public behaviours for sustainable lifestyles. Report 2 of Desk Research commissioned by COI on behalf of DEFRA. Available at: http://collection.europarchive.org/tna/20080530153425/http:/www.sustainabledevelopment.gov.uk/publications/pdf/desk-research2.pdf

Darnton, A., Elster-Jones J., Lucas K., and Brooks M. (2006) *Promoting Pro-Environmental Behaviour: Existing Evidence to Inform Better Policy Making*. London: Department for Environment Food and Rural Affairs (DEFRA).

DEFRA (2005) *Changing Behaviour through Policy Making*. London: Department for Environment Food and Rural Affairs. Available at: www.ukayamut.com/wp-content/uploads/2013/11/defra-4emodel.Pdf

DEFRA (2008) *A Framework for Pro-Environmental Behaviour*. London: Department for Environment Food and Rural Affairs. Available at: www.gov.uk/government/uploads/system/uploads/attachment_data/file/69277/pb13574-behaviours-report-080110.pdf

Durkheim, E. (1950 [1895]) *The Rules of Sociological Method*. Glencoe, IL: The Free Press.

Ferster, C.B. and Skinner, B.F. (1957) *Schedules of Reinforcement*. New York: Appleton-Century-Crofts.

Festinger, L. (1954) An analysis of compliant behavior. In M. Sherif and M. Wilson (eds), *Group Behavior at the Crossroads*. New York: Harper.

Festinger, L. (1957) *A Theory of Cognitive Dissonance*. Palo Alto: Stanford University Press.

Fishbein, M. (1967) Attitude and the prediction of behaviour. In M. Fishbein (ed.), *Readings in Attitude Theory and Measurement*. New York: Wiley.

Fishbein, M. (1968) An investigation of relationships between beliefs about an object and the attitude towards that object. *Human Relationships*, 16, 233–239.

Fishbein, M. and Ajzen, I. (1972) *Beliefs, Attitudes, Intentions and Behaviour: An Introduction to Theory and Research*. Reading, MA: Addison-Wesley.

Fishbein, M. and Ajzen, I. (1974) Attitudes towards objects as predictors of single and multiple behavioural criteria. *Psychological Review*, 81(1), 29–74.

Gausset, Q., Hoff, J., Scheele, C.E., and Nørregaard, E. (2016) Environmental choices: hypocrisy, self-contradictions and the tyranny of everyday life. In J. Hoff and Q. Gausset (eds), *Community Governance and Citizen-driven Initiatives in Climate Change Mitigation*. London: Routledge, 69–88.

Hansen, H. and Andersen, P.N. (2015) Da økologien blev folkekær. *Økologisk*, 29, 14–16. Åbyhøj, Økologisk landsforening. Available at: https://issuu.com/okologidk/docs/magasinet-oekologisk-nr-29

Hoff, J. and Gausset, Q. (eds) (2016) *Community Governance and Citizen-driven Initiatives in Climate Change Mitigation*. London: Routledge.

Jackson, T. (2005) *Motivating Sustainable Consumption. A Review of Evidence on Consumer Behaviour and Behavioural Change*. University of Surrey, Guildford: Centre for Environmental Strategy.

Lucas, K., Brooks, M., Darnton, A., and Elster-Jones, J. (2008) Promoting pro-environmental behaviour: existing evidence and policy implications. *Environmental Science and Policy*, 11(5), 456–466.

Mead, G.H. (1934) *Mind, Self and Society*, edited by C.W. Morris. Chicago: University of Chicago Press.

Norris, P. (2003) *Young People & Political Activism: From the Politics of Loyalties to the Politics of Choice?* Cambridge, MA: Harvard University Press.

North, D.C. (1990) *Institutions, Institutional Change and Economic Performance*. Cambridge: Cambridge University Press.

Økologi (2015) *Økologisk landsforenings historie*. Available at: http://okologi.dk/om-os/organisationen/historie

Parsons, T. (1968 [1937]) *The Structure of Social Action. A Study in Social Theory with Special Reference to a Group of Recent European Writers*. New York, London: Free Press.

Putnam, R.D. (2000) *Bowling Alone: The Collapse and Revival of American Community*. New York: Simon and Schuster.

Shove, E. (2010) Beyond the ABC: climate change policies and theories of social change. *Environment and Planning A*, 42, 1273–1285.

Shove, E., Pantzar, M., and Watson, M. (2012) *The Dynamics of Social Practice. Everyday Life and How it Changes*. London: Sage.

Skinner, B.F. (1938) *Behavior of Organisms*. New York: Appleton-Century-Crofts.

Strobel, B.W., Erichsen, A.C., and Gausset, Q. (2016) The conundrum of calculating carbon footprints. In J. Hoff and Q. Gausset (eds) *Community Governance and Citizen-Driven Initiatives*. London: Routledge.

Thaler, R. and Sustain, C. (2008) *Nudge: Improving Decisions about Health, Wealth and Happiness*. New Haven: Yale University Press.

Van den Pol, J. (2017) Jeg ville ikke kommentere på det, hvis jeg så nogen smide mad ud. Et antropologisk studie af madspil og social dissonans hos klima-og miljøbevidste københavnere. Master thesis, University of Copenhagen, Denmark.

Wapner, P. (1996) *Environmental Activism and World Civic Politics*. Albany: State University of New York Press.

Willer, H. (2017) *European Organic Market Data 2015*. Frick, Switzerland: Research Institute of Organic Agriculture, FiBL. Available at: http://orgprints.org/31200/31/willer-2017-european-data-2015.pdf

Windpower (2016) *Industry statistics*. Available at: www.windpower.org/en/knowledge/statistics/industry_statistics.html

WWF (2014) *Living Planet Report 2014. Species and Spaces, People and Places*. Gland, Switzerland: World Wide Fund for Nature. Available at: http://assets.worldwildlife.org/publications/723/files/original/WWF-LPR2014-low_res.pdf

Part II

Grassroots, green communities and social impact

5 Are vegetables political?

The traces of the Copenhagen Food Coop

Jens Hoff and Mine Islar

Introduction

The Copenhagen Food Coop (CFC), which emerged in 2008, was inspired by the Park Slope Food Coop in New York and driven by the transformative moment created after the Copenhagen Climate Summit (COP15; see Hoff 2016). CFC aims to create an alternative food community in Copenhagen by providing local organic vegetables and fruits to its members at affordable prices, disseminating knowledge on organic and sustainable food production, distribution, and consumption, and collaborating on developing a participatory, inclusive, and transparent community organisation (Tavella and Papadopolous 2017). The cooperative, in 2018, consisted of eight local distribution centres in different neighbourhoods in the city, where members are expected to volunteer three hours a month, primarily ordering and packaging vegetables and fruit. There are also four operational groups in CFC for purchasing, distribution, communication, and events. All members can take part in these groups.

Some scholars see an organisation such as CFC as a part of a broader movement of Alternative Food Networks (AFNs). AFNs have been recognised as a counter-movement to centralised and resource-intensive food systems. They are community organisations that include bottom-up initiatives such as farmers' markets, food cooperatives, and community-supported agriculture (Hinrichs 2003; King 2008; Tavella and Papadopolous 2017). By supplying ecological food and promoting face-to-face contact between producers and consumers, AFNs are said to have a great potential to create environmentally and economically sustainable societies. They also present an alternative arena for transformative change by promoting principles of transparency, inclusiveness, and responsible consumption as opposed to the 'growth at any costs' doctrine that is prevalent in conventional centralised food systems.

In this chapter, we analyse whether it is warranted to see CFC as such an arena for transformative change by exploring CFC in two dimensions. First, we outline the network (political, economic, and cultural) in which CFC is embedded by applying the Actor-Network Theory as established by Bruno Latour (2005) in order to be able to look at the 'ripples in water' effect, or what others would call 'impact'. Second, we analyse these 'ripples' further by applying theories on political community and new network politics in order to investigate the political character of CFC. Despite the large diversity of scholarly studies on the development of local movements and cooperativism, little has been said about the co-articulation of collective practices and their political significance.

Some scholars argue that localised alternative food networks, which are also at the centre of movements such as Transition Towns, might create unreflexive localism by 'depoliticising' the local (DuPuis and Goodman 2005). Through the case of the CFC, we demonstrate that these networks are indeed political. Their political character might be difficult to detect if one focuses only on the traditional political system and traditional politics (mode 1 politics according to Bang 2009, see also de Vries 2007), especially as some of these movements, including the CFC, explicitly declare themselves 'apolitical'. However, their political character becomes much more obvious if one focuses on issues, political community, and the creation of political capital (mode 2 politics according to Bang 2009, see also Latour 2007).

Theory and method

Theory

The de-politicisation argument takes its point of departure in post-foundational political theory, especially as formulated by Chantal Mouffe (2005a, 2005b). Mouffe distinguishes between 'politics' and 'the political', where 'politics' is about 'the set of practices and institutions through which order is created', while 'the political' refers to 'the dimension of antagonism' (Mouffe 2005a, 9). Thus, politics refer to the social sphere where 'traditional politics' take place, such as parliament and government. In contrast, 'the political' refers to the order of representations through which society is given meaning, or discourses that can manifest themselves outside the sphere of politics. More precisely, the political is that order of discourse that acknowledges the existence of power, conflict, division, and contingency in society.

For Mouffe, a discourse is post-political when it misrecognises the constructed and therefore contingent nature of the social, when it conceals the fact that each construction entails certain exclusions and can therefore generate conflicts or antagonisms, and when it obscures the fact that the construction of the social inevitably entails acts of power. Moreover, according to Zizek (1999), the effect of such post-political representations is that they exclude the possibility of more far-reaching social transformations. Democracy itself is also at stake in the post-political, as democracy according to these scholars is impossible when conflict, power, or exclusion remain invisible and uncontestable (Kenis and Mathjis 2014, 174).

The scholar Erik Swyngedouw demonstrates that post-politics manifests itself, in particular, in many of the currently predominant environmental discourses (Swyngedouw 2007, 2010, 2013). Following this notion, Kenis and Mathjis (2014, 174) state in their work on the Transition Town movement that the pivotal question is therefore whether and to what extent such alternative environmental movements succeed in going beyond post-politics. Specially, the questions at hand is how recent localisation discourses, as put forward by for example Transition Towns, relate to the problem of post-politics.

Recognising that several scholars, e.g. North (2010) and Urry (2011), have characterised Transition Towns as 'deeply' or 'significantly' political, Kenis and Mathjis discuss the pros and cons of local food provision, which is a central concern in the movement. They recognise that there are environmental advantages of local food production, such as decreases in energy and pollution costs related to global transportation and distribution systems and to often more sustainable productions methods such as organic farming. However, they also show how the 'local' is often idealised and presented as radical and subversive, while the global is conceived as hegemonic and oppressive. Indeed, as several authors have shown (Allen 2004; Campbell 2004, and others), localisation risks being associated with forms of protectionism, particularism, patriotism, conservatism, and even xenophobia and also risks becoming a 'defensive localism'. They therefore conclude that a local food system is not political or post-political in itself; it depends on how it is given meaning discursively, and in particular whether power and conflict are rendered visible or not in the localisation discourse (Kenis and Mathjis 2014, 176).

In their critique of the Transition Town movement, Kenis and Mathjis (2014, 176) argue that:

> Rather than being a new grassroots environmental movement that tries to tackle the twin problems of climate change and peak oil through building resilient local communities ... it appears as if Transition Towns is first and foremost a localization movement, which refers to climate change and peak oil to reinforce its case.

Crucial in this regard, they say, 'is that not only the "local" but also associated terms such as "community" and "rurality" are socially constructed in such a way as to become strongly depoliticized' (Kenis and Mathjis 2014, 176).

While we do not want to imply that Kenis and Mathjis' analysis of the Transition Town movement is wrong or invalid, we suggest in our analysis below that their conclusion is, to a very large extent, coloured by their theoretical point of departure in post-foundational political analysis. This type of analysis is, in our view, strangely conservative in its view on politics as being centred on traditional political institutions and as the political expressing itself only as visible political struggles against an oppressor. Following Henrik P. Bang (2009), we call this focus 'mode 1 politics'. What we highlight here is the type of politics conducted in and through movements such as CFC (and Transition Towns), and which focuses on practical work, collective decision-making, identity formation, and the production of political capital, or what Bang (2009) calls 'mode

2 politics', which constitutes 'bringing political community back in' (Bang 2009, 105).[1] This way of analysing 'sub-politics', as Beck (1996, 2006) calls it, is very much in line with a Science and Technology Studies (STS) perspective on politics as a praxis, or an accepted practice or custom which constitutes a political object ('a common good') or formulates an issue around which a public can be formed (Marres 2005; Latour 2007; de Vries 2007; see also the chapter by Christensen *et al.* in this volume).

According to Bang, the malaise that characterises modern-day politics is that the democratic chain between political authorities and laypeople is broken, and that there is a neglect of the role of laypeople in their political communities caused by the dominant political elite. Thus, one has to appreciate that traditional parliamentary politics and communal action express two different modes of 'the political', which Bang terms *politics-policy* and *policy-politics* (Bang 2009, 102).

In its *politics-policy* mode, the political revolves around the question of how demands are converted into collective decisions, and the core question for democracy here is how to ensure that conflicting interests and identities can acquire free and equal access to, and recognition in, the political decision-making process (mode 1 politics).

In its *policy-politics* mode, the political concerns the question of how collective decisions are articulated and programmed as policies and then delivered to the people. Here the issue for democracy is how to establish communicative and interactive linkages between political authorities and laypeople in the political community for the sake of balancing existing asymmetries of power in and through concrete political actions (mode 2 politics).

This distinction implies that the actions that some political science scholars (Stoker 2006; and indeed Mouffe 2005a, 2005b; and Kenis and Mathjis 2014) write off as uncritical and inconsequential with regard to mode 1 politics – which include increasing citizen involvement in network governance, public–private–voluntary partnerships, new forms of participation such as alternative food networks, and new forms of policy deliberation – may prove highly significant and relevant to the conduct of good governance in mode 2 politics. Examples of such citizen involvement are the *everyday makers* and the *expert citizens* described by Bang and Sørensen (1999, 2001), that is, people who think in terms of 'making a difference', rather than participating in mainstream political decision-making. The important thing here, as Bang (2009, 104) stresses, is that everyday makers and expert citizens:

> typically make their choices as part of a project identity, rather than an oppositional or legitimating one. To them, participation is not a matter of being either *for* or *against* 'the system'. They adopt an oppositional or legitimating identity to the extent that it is necessary to developing their project identity and, as such, engage, or do not engage, in policy articulation and delivery, depending on their concerns at a particular time.
>
> (Emphases added)

According to this line of reasoning, arguing, as for example Kenis and Mathjis do, that not having or not expressing an oppositional identity is the same as being apolitical misses the point. Actors who are 'apolitical' when it comes to mode 1 politics might very well be active in mode 2 politics. As the analysis below suggests, this is very much the case with members of the CFC, even though both the organisation as such as well as its individual members have in fact also been involved to some extent in mode 1 politics. So instead of being apolitical, the organisation might in fact qualify as being what one could call 'double political'.

In order to further qualify the character of democratic political community, we also have to realise, according to Bang (2003, cf. Easton 1965), that a community such as this does not emanate from trust in authorities or legitimation of the political regime. Such a community is the result of the empowering of lay-people as capable and knowledgeable members of a political community who share a common division of labour and, in the course of time, develop a mutual identification that springs from their concrete experiences with how to 'make a difference' (their praxis). We should also realise that what is important in such a community is not only the kind of social capital built by networking and trusting one another, but also, and maybe more consequential, the kind of political capital that comes from communicating and interacting in the interest of solving common concerns.

In the analysis below, we demonstrate how mode 1 and mode 2 politics are expressed in CFC. We deal with mode 1 politics in relation to an analysis of the 'inner' network of CFC, or how the organisation as such has developed over time. Mode 2 politics is dealt with in relation to an analysis of the wider network of CFC, or the traces CFC has left in the 'glocal' environment. We divide the discussion in this way in order to sharpen the analytical lens because both types of politics are clearly present in both networks. Our focus points are those emphasised by political community theory: the common work carried out, the decision-making processes around the work, the project identity developed, and the political capital built. However, before proceeding with the analysis we take a look at the method used in our study of CFC.

Method

We are inspired methodologically by the Actor-Network Theory (ANT) because we want to describe and understand the traces left behind by the CFC, how its objects and practices bring new realities into being, and how it thereby produces power and identity relations. However, we are well aware that what we are producing is not a 'real' ANT-analysis, as we have already chosen 'the political' as our topos of investigation, thereby excluding other ways of looking at the CFC (see Mol 1999). Therefore, we have to some extent already answered the question posed in the title of the chapter: 'Are vegetables political' with a yes, which forecloses the possibility of looking at the vegetables bought and distributed by CFC as something else – for example, as artefacts producing community, locality, or simply health. That we are here deliberately choosing one option among others is neatly summed up by one of our key informants, who says: 'From the very beginning the word was "ecology for everyone", it has not been about politics. Vegetables do not have to be political' (IP4, 9). What we discuss below is therefore not whether vegetables are political or not, but rather in what way they are political and the possible effects of this politicisation. We therefore cannot redeem the demands of a 'real' ANT-analysis as that of starting with a 'flat ontology', or 'keeping the social (completely) flat', as Latour (2005, 165ff) puts it. But even though we deviate from ANT in this respect (and possibly many others), we follow many of the recommendations laid down in ANT, and prey on the ANT vocabulary, thereby producing some kind of 'bastardised ANT' science.

Our analysis therefore resembles an ANT analysis in many ways. We use a 'reassembling technique' for looking at networks. We do not take the networks or their character as a given, but look at how they are constructed and what keeps them together. By doing so, we focus on all actants involved, human as well as non-human, and pay a good deal of attention to the role of non-human actants (material objects) such as, for example, the organic food bought and distributed, the food distribution hubs, the information technology used to organise it all, and the written principles of the organisation. In this sense, we are true to a basic tenet in ANT, which is '*recording* not filtering out, *describing* not disciplining' (Latour 2005, 55, emphases added).

Second, when we analyse the networks, we avoid 'essentialising' them, which means that we allow ourselves to see them as multiple, as being open to different interpretations about their purpose, their history, and the new realities they have created. This is what ANT terms seeing the world as 'matters of concern' rather than as 'matters of fact' (Latour 2005, 116). This is understanding that there can be different 'stories' about the CFC and its (lack of) societal importance (Papazu 2016). However, as Latour emphasises, there cannot be an indefinite number of stories. Multiplicity is not the same as deconstruction, and the presentation of different realities must always have an empirical grounding. (Latour 2005, 120).

Third, as suggested by ANT, we will also 'feed off controversies'. Our focus is the narrow controversies concerning the political/non-political character of the networks we are studying, and we will try to examine these controversies in a way that keeps them open, as matters of concern or assemblages rather than as matters of fact.

Finally, we employ the same definition of a network as that used in ANT. Latour (2005, 132) defines a network by four important features: (1) it is a point-to-point connection, which is physically traceable and can thus be recorded empirically; (2) such a connection leaves empty most of what is not connected; (3) this connection is not made for free, it requires a (research) effort; and (4) a network is not made of nylon thread, words, or any durable substance, but is a *trace* left behind by some moving agent (thus the word 'traces' in the title of this chapter). The second point is very important here: it directs our attention to the fact that there are many things about the CFC that we do not know and never will know. Our and others' stories about the CFC and its importance in the world are only a fraction of what has been told and what could be told, but we insist, while recognising all the limitations that lie in our data and our empirical work, that our story about the politics of CFC is as important as many others.

Data

Our analysis of the CFC is a single-case study based on the assumption that the CFC is an extreme or critical case of an alternative food network. As such, it might represent a new form of environmental politics, which some have called do-it-yourself activism (DIY) and others have called simply 'practivism'. We argue that CFC is a particularly vivid and information-rich example of environmental practivism, in this case of alternative food provisioning practices and networks. It is information-rich because it:

- has existed for ten years;
- has experienced significant growth and manifest crises (3000 members in 2012, now down to 500–600 but still with eight distribution centres in Copenhagen);
- represents a radical ecological movement (according to Saunders 2013) with explicitly stated goals and basic principles relating to transformed relations with production, distribution, and provisioning of food; and
- has a remarkable number of traces in the form of connections with food suppliers, environmental organisations, political parties, local cultural institutions, small start-ups, huge supermarket chains, the media, etc.

The data we have collected consist of seven interviews (of eight people) with members or former members of CFC and with people in the network surrounding CFC, such as a politician, a journalist, a person from a small commercial spin-off, and a Corporate Social Responsibility (CSR) director from a major supermarket chain. All interviews were conducted between 29 November 2017 and 14 February 2018.[2] We adopted purposeful sampling in choosing the informants for our interviews. This is a technique widely used in qualitative research for the identification and selection of information-rich cases for the most effective use of materials (Patton 2002). This involves identifying and selecting individuals or groups of individuals that are especially knowledgeable about or experienced with CFC and local food economy or policies.

Furthermore, we have studied a number of texts concerning CFC, such as a book written by some its founders (Lloyd and Pass 2013), two master's theses written about different aspects of CFC (Christensen 2016; la Cour 2017), which contain transcripts of interviews, and we have informed ourselves on the CFC's website (www.kbhff.dk). Finally, we have conducted a small media analysis of the mentions of CFC in the mainstream media.

The everyday politics of CFC: reassembling the networks

In this section we engage in an analysis of the concrete practices of CFC that weave its networks together and thereby constitutes CFC into a cohesive organisation. As our focus is on the practices constituting mode 2 politics, we concern ourselves especially with (a) the common division of labour, which in this case is constituted by the buying and selling of organic food; (b) the mode of decision-making used to frame this activity; (c) the mutual identification or project identity that springs from these experiences; and (d) the political capital developed by the involved individuals (learning 'how to make a difference'). We shall deal with these four points in turn.

The alternative economy of food: buying and selling organic food

The start-up

The inspiration for the Copenhagen Food Coop came from the Park Slope Food Coop in Brooklyn, New York. Park Slope Food Coop is a grocery store run by its 20,000 members, who each volunteer to work in the store for 2.5 hours a month. A founding member of CFC talks about how it all began:

> A guy called Morten came back from New York and said: 'Damn it, we will make a similar thing in Copenhagen'.... The first couple of years people spent a very long time talking about what kind of shop they wanted. But there were no groceries, there was no plan, and there was no money, so it was just empty talk. At last some people got impatient and said: 'Now we just buy some vegetables and see what happens'. They made an agreement with a person living on Svanholm [a huge ecological farm 50 km from Copenhagen], who was going to Copenhagen on Wednesdays because she took African dance classes and could transport some root vegetables in her car.
>
> (IP4, 1)

This rather loose and coincidental way of organising the buying and selling of the vegetables and fruit did not work very well, so a couple of people stepped in to organise things better:

> Me and someone named Kristine were rather frustrated about this good idea not being handled very well, so we asked: 'What about starting up properly after the summer holiday?' This was in the spring of 2009. So we went out and made an agreement with some people at Nørrebro [Korsgade], and we were allowed to use their basement. We also rewrote the original manifesto, together with a woman called Maiken.
>
> (IP4, 2)

In order to build up the organisation, the organising group used online tools quite extensively. They called a start-up meeting in August 2009 at their location in Nørrebro and they sent an invitation/the manifesto to around 250 people who were on their mailing list at that time. Then they created an organisation website and a wiki (a user-driven website) where people could sign up for three-hour work slots as volunteers, and finally they created a Google sheet to administer orders for the vegetables.

What happened next?

> Thirty people turned up in that basement in Korsgade and we told them: 'In two weeks we will have the first vegetables, but this will demand two things: First, that you become members of the organisation and that you pay for a bag [of vegetables]. Second, that you pay in advance and are guaranteed a bag of vegetables worth 100 DKK'. We divided ourselves into three groups: a shop group, a procurement group, and a communications group. And we were lucky – a friend of my ex-girlfriend knew a local ecological farmer, who owns Kieselgaard, a farm near Slagelse [a city on Sealand]. He would send some vegetables if we paid for the freight.
>
> (IP4, 3)

This organisation of the food coop worked well for a long time: there were very few Wednesdays where there were no vegetables. After a year, CFC had 500 members, and decided to become more publicly known. Whether due to the work of the communication group or public interest, articles about CFC began to appear in several daily newspapers, and the coop was featured on television (TV2 Lorry, a local Copenhagen TV station). Following this, CFC enrolled around 50 new members weekly. The original basement in Korsgade became too small, and members began to start their own local division of the coop, first in Amager, then in Vesterbro, and on to different other parts of Copenhagen. During 2011–2012, the coop increased to 10–12 local divisions.

> It was completely crazy to create an organisation based on voluntary manpower only, where all tasks have to be broken down into three-hour work slots per month for everyone, especially if you have several thousand people who are loosely coupled – they sign up when they feel like it. They may or may not read the newsletter. It was complete chaos. And simultaneously we had decentralised the economy of the coop.
>
> (IP4, 3)

This situation resulted in a huge crisis in 2013, after which the coop was reorganised (see below). During this organisational turmoil, the number of bags of vegetables sold fell from 1200 per week to around 500–600. The number of members also declined from probably around 3000 (other sources have 5000 members) to around 1300 (IP1, 20; IP2 + 3, 16). (We address the restructuring of CFC after 2013 in greater detail below.)

We finish this section by describing the relationship of CFC to its suppliers, how prices are decided on, and the societal and political dimension of these factors. The fundamentals of these relationships are depicted in the ten guiding principles in the CFC manifesto, which was written in 2009. These principles are:

- We buy and distribute food which is grown and produced ecologically [following ecological, that is, also biodynamical principles].
- [The food we buy and distribute] is grown as locally as it is practically possible.
- [The food we buy and distribute] will mirror the seasons.
- [We distribute food in a way that] supports fair and direct trade.
- [We distribute food in a way that] is environmentally friendly.
- [We distribute food in a way that] communicates and promotes knowledge about food and ecology.
- [We distribute food in a way that] makes us [the CFC] economically sustainable and independent.
- [We distribute food in a way that] is transparent and creates confidence in all parts of the production- and distribution-cycle.
- [We distribute food in a way that] is close to the members and affordable. Ecological food should not be an expensive luxury.
- CFC shall be run by a local working community.

These principles clearly demonstrate that the CFC is not satisfied with the traditional (non-ecological) way of producing food in Denmark and the way the food is distributed.[3] In both areas, the CFC aims to make a difference by supporting fair and direct trade and by growing the produce as locally as practically possible. In practice, supporting fair and direct trade has meant that the CFC has seldom bargained very hard with producers about their prices as long as they have been considered 'fair', and the vegetables are affordable for the CFC members. Direct trade has also meant that the CFC has created direct, personal contact with farmers wherever this has been possible. This more direct relationship between the food producers and CFC eliminates unnecessary distribution links, and also encourages farmers to continue their ecological food production. The direct, personal contact with farmers was made easier by the fact that some of the people who started CFC had worked in the National Association for Ecological Farmers (*Økologisk Landsforening*), and therefore knew many of the farmers already (IP1, 10).

We also note that from the very beginning CFC was seen as a platform for more than just buying and distributing good and inexpensive food. The founding CFC members also intended that it should be a platform for local collaboration, and that one should be able to find others who were interested in, for example, starting an urban garden project or buying ecological diapers. 'So in this way we saw some projects coming out of CFC like the urban garden on Nørrebro [Byhaven 2200], an apple plantation and apple juice producer [Farendløse Mosteri], and a restaurant [Blå Congo]' (IP4, 8).

The networks

It is notoriously difficult to describe networks, especially in writing. From which vantage point should one describe the network? What are the types of connections that constitute the network, and how does one account for them? And should one describe all types of connections, even the ones that are very irregular? These questions are very pertinent because the CFC consists of networks within networks. Thus, in terms of describing the overall assemblage of CFC, and CFC's relationships to other actors, we have relied heavily on 3–4 key informants who are able to describe 'the big picture'[4] of CFC as an organisation as it looks from their perspective. These are typically people who have been members of 'kollektivgruppen', later called 'the board', or people who have been doing paid administrative work for CFC. We let these people speak on behalf of the organisation knowing that while they also may not have the total picture, their sense of CFC could be valuable to our analysis. In terms of describing the types of connections that constitute this network between organisations in which CFC is embedded, our strategy has been to record all the types of connections that are mentioned in our interviews. We describe these connections below in a number of displays grouped by type of actors; the displays also briefly describe the content of the connection.

The CFC is also constituted by all the local divisions that constitute networks in their own right, and who create another type of network that together 'performs' the whole organisation. These local networks, in which the local distribution hubs (like the basement in Korsgade) play a very important role as a 'plug-in' (Latour 2005, 207–210) that connects everyone, are not described here because our focus is to describe and understand the wider societal traces of the CFC. Also, we only describe the analogue, or face-to-face, connections here, leaving the description of the online network and presence of CFC for another analysis. However, we know that CFC's online presence is an important part of the entire network. For example, even though the CFC has its own Facebook page and a group called 'fællesgrupper', or 'community groups', there are at least another 180 Facebook profiles created around CFC (IP1, 12).

First, the *economic component* of the CFC's network consists of the suppliers or the ecological farms from which CFC buy their products. We know that Svanholm (and also Kieselgården, cf. above) was the first supplier, and that the next was Birkemosegård (both are now well-known ecological farms). CFC has also made a transportation arrangement with Birkemosegård, and also has so-called 'hatching patrols', that is, CFC members visiting and working at the farm. The economic component also consists of market-oriented spin-offs where members of the CFC have created their own enterprises. These are outfits such as Farendløse Mosteri (the aforementioned apple plantation and juice producer) and the restaurant Blå Congo.[5]

The economic component also consists of larger commercial companies which have used CFC as an inspiration for their own business or who have recruited members from the CFC to work for them. For example, Årstiderne is a distribution concept whereby consumers can order weekly boxes of food on a website and have them delivered to their doorstep, a concept that has turned out to be hugely successful. And the big supermarket chain COOP, which owns a chain of supermarkets named Fakta, DagligBrugsen and Irma, has studied CFC intensively and used it as an inspiration for a renewal of their business in a more ecological direction (IP8).

The economic component of the CFC network is summarised in Table 5.1.

The second CFC network component is the political network, which is related to mode 1 politics. This component is described in detail below.

The third component of CFC's network is what we call the 'food culture or lifestyle' component. It is immense: it comprises numerous actors of various kinds such as schools, colleges, and a huge amount of temporary engagements. Examples of such short-term but highly visible and effective engagements are the food stand that CFC used to host at Roskilde-festivalen, a huge annual one-week annual music festival held near the city of Roskilde, which attracts foreign guests from like-minded food cooperatives in, for example, Poland or South Korea.

There are two dimensions of the food culture and lifestyle network that are especially interesting. The first is the engagement with local schools. Some local schools early in CFC's history offered to house the local CFC distribution hub. This led to different types of collaborations between the local division of CFC and the schools, such as the CFC delivering vegetables to the school canteens and local CFCs cooking meals for teachers and/or parents at staff meetings in exchange for rent. Also, members of CFC have given talks about ecological food to both parents and teachers. This collaboration between local divisions of CFC and local schools has given CFCs a very good grounding in their locality, and also demonstrates one of Latour's points: that the materiality or non-human actor (here the school buildings), plays a very important role in performing the local network.

The second interesting dimension of the food and lifestyle network is the many foreigners, especially from Anglo-Saxon and Eastern European countries,

Table 5.1 CFC's economic network (partial list)

Entity	Type of connection
Svanholm (farm)	Supplier
Birkemosegård (farm)	Supplier
Bellingehus (farm)	Supplier
KysØko (farm)	Supplier
Farendløse Mosteri (apple juice producer)	Spin-off
Fejø Æbleplantage (apple plantation)	Supplier, spin-off
Blå Congo (restaurant)	Spin-off
Årstiderne (food supplier)	Inspiration from CFC
COOP (supermarket chain)	Inspiration from CFC
LØS Market (packaging free shop)	Buyer
Karise Permatopia (ecovillage)	Spin-off

who have joined the network in order to learn Danish and meet Danes. This has made the CFC into a kind of 'integration hub' and nicely illustrates how the local is (also) globally constructed. In line with this are the many visitors from food cooperatives abroad that the CFC has received. The food culture and life-style part of the network is summarised in Table 5.2.

To assess CFC's relationship with the media, we made a quick InfoMedia search in the major daily Danish newspapers from 2009 onward (search term <*Københavns Fødevarefællesskab*>). This yielded a total of 10 articles about CFC. Most of these articles were positive, but they also reflected the political outlook of the newspapers. The centre-left newspapers tended to deal with the CFC as part of 'the great transition', whereas the centre-right newspapers tended to deal with CFC as more of a lifestyle trend or food fashion, even combining the description of CFC with recipes.

Table 5.3 summarises the media component of the network.

The mode(s) of decision-making

According to Bang (2009), two types of participation are necessary for building political capital. One is *strategic communication* with a political authority in

Table 5.2 CFC's cultural network (partial list)

Entity	Type of connection
Ingerslevsgade Skole (primary school)	Cooked meals in exchange for housing
Amager Fælled Skole (primary school)	Housing
Metropol (college)	Presentation for
Københavns Madhus (municipal catering)	Inspiration for
Byhaven 2200 (city garden)	Inspiration for
Roskilde Festival (music festival)	Co-produced food at
Folkets Hus på Nørrebro (local citizen's hall)	Collaboration with
Miljøpunkt Østerbro (local Agenda21 office)	Collaboration with
Frederiksberg Grønne Dage	Event at
Østerbro Weekend	Cooked meals
Copenhagen Cooking Festival	Cooked meals

Table 5.3 CFC's media network (partial list)

Entity	Type of connection
Information (daily newspaper)	5 articles on CFC
Weekendavisen (weekly newspaper)	2 articles on CFC
Politiken (daily newspaper)	1 article on CFC
Jyllandsposten	1 article on CFC
Berlingske Tidende	1 article on CFC
TV2 Lorry (local Copenhagen TV channel)	Appearance
Internet, social media	Not analysed

order to influence, articulate, and assess politics and policy. The second is *conventional communication*, where members of a community become engaged, discuss values, and deliberate on what could be done to build a reflexive political community. Like many cooperatives, CFC's decision-making structure is along the lines of conventional communication, which is operationalised through participatory and consensual decision-making processes. However, the rapid expansion of CFC with an increasing number of members as well as the economic hardship the organisation faced in 2013 led to some difficulties with this type of participatory model and organisational structure:

> We did not have any visible leadership (before 2013), so no one felt that they really had a mandate to make decisions.
>
> (IP4, 5)

> Consensus in member meetings gave problems because you must be able to make decisions.... We didn't have a board from the start. Instead, we had a group called 'the collective group', who were representatives of the different working groups that met and talked about what was important to do. This group was always open, so everyone could help and take responsibility.
>
> (IP4, 6)

In order to improve the economic situation and to overcome ambiguities about who is responsible for what, CFC underwent a structural re-organisation in 2013, which resulted in three clearly distinguishable levels of responsibility: the coordination group, operational groups, and the local divisions. The board or the coordination group is at the top of CFC's organisation structure: it meets at least once a month to ensure that financial and legal responsibilities are monitored. The most important decisions about CFC's operation are taken by the four operational groups noted above: purchasing, distribution, communication, and events. All members can take part in these groups. The operational groups hold coordinating meetings every other month. Most local divisions have also started to follow a team structure in which members share the different responsibilities.

CFC members do not always share a common strategic focus, particularly when it comes to long-term planning and deciding on options concerning future development (Tavella and Papadopoulos 2017). For instance, at the general assembly for discussing a new business plan in 2017, one of the suggestions of the distribution group was to set a higher food price for those members who could not spare three hours a month to volunteer. Several members explicitly objected to this proposal and emphasised the importance of the cooperative model of CFC – it is not only about ecological food, it is also a community (*fællesskab*).

Project identity

Most members of the CFC have a very clearly developed idea of what Bang (2009), following Castells (1997), calls a project identity. A project identity is formed around the identification of the participants with the project or the 'course' itself. It is very context-dependent: whether members will fight against or collaborate with authorities depends on the specific situation. Also, as in identity politics, identity formation is very much a project, the realisation of which depends crucially on the transformative capacity of oneself and others (Bang 2003). This is important here because, as we shall see, the identity of CFC members seems to change over time.

The project or the 'course' is formulated very clearly in the aforementioned CFC manifesto; most members seem to know the ten guiding principles in the CFC manifesto quite well and use them as an inspiration for their actions. That the activism in CFC is oriented towards a project is also very clear from the following quote: 'It is more a "course"; a change that you want to see and one which you create very concretely' (IP2 + 3, 14).

However, even though it is important for most members to make a difference through their concrete actions in CFC, some seem to focus more on being in a community with other like-minded people than on the food component of the community. CFC member Jonas says: 'For me it has all been very much about the community' (IP1, 1). But recognising that the core activities in CFC concern the buying and distribution of food, he also says: 'There is no community without the food, and there is no food without the community' (IP1, 19). For others, the project's political implications were clearly the most important factor for them from the beginning, but after some time the community starts to become important as well. Louise, for example, says: 'Now the community is beginning to become really important to me.... It is a social and cosy thing' (IP2 + 3, 3). The community aspect is also stressed by the fact that most local divisions arrange occasional or regularly scheduled common dinners that may include 15–20 people.

A small membership survey (140 respondents) conducted in 2014 assessed what motivates people to become members of the CFC. The survey results confirm that project identity is the most important factor for members: only 10 per cent said that for them the political statement is the most important, and the remaining 90 per cent said their primary motivation is working with the vegetables, high quality food, and the community created around it.

However, this survey as well as an earlier survey also showed that a significant number of members might never reach the state where they really experience the cosy community. There is a high turnover in the organisation: 50 per cent of the members have been members for less than a year, while 25 per cent have been members for four years or more. This suggests the existence of an A- and a B-team within the organisation, or perhaps between what Bang (2009) calls 'expert citizens' and 'everyday makers'. While there is little doubt that the majority of CFC members qualify as 'everyday makers' according to Bang's criteria,[6] it is more doubtful whether the minority of long-term members or people elected to the board or the few in paid positions qualify as 'expert citizens'. By definition an 'expert citizen' is:

> most often a (new) professional, ... who feels that he/she can articulate and implement policy as well as, and even better than, politicians and other professionals from the public and private domain.... Expert citizens put policy before politics in their project identity. They are more concerned with having an impact on the concrete articulation and delivery of policies that helps them in realizing their various projects than in fighting the authorities....
>
> (Bang 2009, 131)

One of the reasons that this definition might not quite fit the CFC is the composition of its membership. The organisation is composed mainly of students, people on pensions (the elderly), and the unemployed. Also, as we saw, there is a substantial number of people of foreign origin in CFC that overlaps with the three membership groups. The expert citizens in the organisation might therefore be somewhat hidden but are surely there working to (also) influence the political world with their ideas. Likewise, there are also everyday makers rising to a certain status in the organisation due to their experience and long service in CFC. These members, mostly elderly women, are called 'vegetable queens'. While they contribute to the stability and continuity of the organisation, they can also be somewhat inflexible and difficult to co-work with (IP2 + 3, 12).

Development of political capital

In traditional political science, the concept of political capital is often operationalised through the concepts of internal and external efficacy. Internal efficacy refers to a person's self-confidence in relation to politics; external efficacy refers to a person's perception of how much he/she is able to influence politics at different levels (i.e. local, national). This conception of political capital is very much related to politics in mode 1. As we are focusing on politics in mode 2 here, we use Bang's (2003) perception of political capital instead, which is concerned with what people learn from communicating and interacting in order to solve common concerns.

In CFC, the individual learning seems to concern values and skills as well as certain ways of communicating to/with other members of the CFC.

Values and skills

One of the important things about CFC is that there seems to be a low threshold for entering the organisation and learning about sustainability and vegetables (IP2+3, 6). This is stressed by a journalist we interviewed who has followed green parties, movements and initiatives in Denmark over many years. He says:

I think there is a huge value in the formative experience, where the participants are taught some new values through these communities, who have explicitly stated goals. They build self-confidence and a belief that it is actually possible to do something.

(IP6, 8)

One of the 'founding fathers' of CFC disagrees with the journalist on the question of values and stresses the teaching of concrete skills that goes on in the organisation:

The values were there already. What we have done is to provide a platform for the participants to actually live out these values. Both in the sense that they learn something about from where their food comes, what ecology means, what the difference between ecology and biodynamic is.... To eat root crops all year around, and find out what is in season at any given time. This is a form of 'green formation'... but most important of all I think, is that people have learned to be part of a community – maybe caught a glimpse of how a sustainable future might look.

(IP4, 11)

So while there is some disagreement on whether the members of the CFC already have green values before entering the organisation or whether this is something they learn through their membership, there is more agreement on the skills dimension – that members actually learn about sustainability, ecology, and other environmental issues. This learning has resulted in members being asked by supermarket chains and other commercial vendors to give presentations about CFC to them.

Communication

One of the essential dimensions of building political capital is communication. The mode 2 politics approach (Bang 2009) stresses the importance of a new, more communicative and interactive democratic model. This model opens up the possibility of a fusion of identity politics and project politics and aims to get things done in a prudent manner by establishing balanced and discursive two-way relations of autonomy and dependence between political authorities and laypeople, but also between laypeople themselves (Bang 2009, 117).

Thus, in order to get things done in a prudent manner, it is necessary to learn to communicate in an inclusive and reciprocal manner, which is both convincing and creates ownership of the project. This is an important skill that at least some members of the CFC have learned:

And this is something I have learned from CFC: how to talk to people so that they feel that they can become a part of it [the CFC]. It is about not talking negatively about anybody in order to put yourself in the spotlight. It is about communicating solely in a positive and vulnerable way. The positive thing is about only talking about what we want to do, and not referring to what we think others do wrong … because it has to be so that the idea and the project you propose must be strong enough to convince people that this is the way to go.

(IP7, 7)

The inner network and politics in mode 1

We explore in this section the inner network of CFC, and the 'conflictual' incidents that highlight mode 1 politics, or what ANT would call 'controversies'. Our analysis of the controversies in CFC's inner network and its cooperation with political parties through some of the individual members indicates the extent to which CFC is involved in mode 1 politics. CFCs explicitly declare themselves 'apolitical' (IP4, 9). However, the organisation's evolution has been engaged with and affected by emerging green parties such as Alternative (*Alternativet*) as well as two incidents, the 'avocado incident' and the 'pea-sprout incident', that led non-parliamentarian radical left-wingers to withdraw from CFC in its early years.

The avocado incident refers to a contentious moment in CFC's inner network when some members wanted to boycott organic avocados that were imported from Israel by the organic company Solhjulet during the winter season of 2009–2010. CFC wanted to offer more than locally produced food that winter because root crops are the only vegetables one can get in the winter in Denmark. They made an agreement with Solhjulet to expand their winter inventory. However, bringing in avocados from Israel created an uproar in the CFC, which was tied to the more general boycott-Israel movement of the time, and CFC reconsidered the principles they apply when buying food. In order to manage different expectations of their members, they held a grand meeting where a draft of the 10 basic principles was presented. These principles, which were formulated in line with food communities in, for example, England, are still at the core of the food community (see above).

> We elegantly avoided saying that we boycotted Israel, and if we buy the food locally it is practically possible. And then we have never bought goods from abroad again – except olive oil.
>
> (IP4, 8)

The second controversy, which we refer to as the pea-sprout incident, concerns a conflict that arose when CFC's local shop in the so-called Youth House got a delivery of pea-sprouts from an open prison horticulture:

> There was an ultimatum from the Youth House that we could not sell goods that were grown in a prison. We didn't hurry to make a decision, and we took a debate meeting where we talked about what is really up and down in

this. What are the conditions under which these prisoners work, and what is the perspective for them, etc.? I was not alone, but I heard that there were really good arguments for buying goods from open prisons as it is also a way to support a rehabilitation process. The Youth House's main argument was that they did not like prisons and were in principle opposed to this idea.

(IP4, 8)

So, in that way we have broken with the very leftist identity and have instead been much more middle-seeking in the way we have talked about ourselves. Where the word from the beginning has been ecology for all, and it's not about politics but about vegetables. Vegetables do not have to be political, we say we will be part of a sustainable future but there's no party politics in it, though the Red/Green Alliance [*Enhedslisten*] and others have of course backed up much of what we've been doing.

(IP4, 9)

Apart from this, there are many links and overlap of people between CFC and the Red/Green party (*Enhedslisten*), the Socialist People's Party (*Socialistisk Folkeparti*), and the Alternative (*Alternativet*) which indirectly expand CFC's influence on the political agenda of these parties (Table 5.4). For example, one of the members of Alternative, who is on the Copenhagen Municipal Council, stresses that the food policy of the local division of the party has been much influenced by the CFC (IP5, 4).

Conclusion

Vegetables are indeed political, and they are even political in different ways. They are political in what we called mode 1 as well as in mode 2 politics. But they need not be. For some people, CFC's vegetables can be seen as synonymous with community, locality, or health, which means they are non-political. In that sense, some of our key informants are right: there are many stories about the CFC and only a few seem to centre on politics.

Table 5.4 CFC's political network

Entity	Type of connection
The Alternative Party	Person overlap, flow of ideas
Red/Green Alliance	Person overlap, flow of ideas
Peoples Socialist Party (SF)	Person overlap, flow of ideas
NOAH (Friends of the Earth, Denmark))	Common campaign
Local Council (Vesterbro)	Contacted CFC for representation
Friends of Amager Fælled (grassroot org.)	Person overlap
Økologisk Landsforening (National Association for Ecological Farming)	Support for

However, what we have chosen to tell here is the story of CFC as a political topos. Central in this story is that we see CFC as political in the mode 1 sense, as it has had visible controversies and conflicts within the organisation that resulted in a radical non-parliamentarian left-wing group leaving the organisation. It is also political in the mode 1 sense in that it has person overlap with established green and/or left-wing parties and because the ideas and actions of CFC have influenced these parties.

CFC is also political in mode 2 because its core activity is communal action around buying and selling organic vegetables. In doing so, it clearly attempts to make a difference both to the traditional (non-ecological) way of producing food and to the way food is distributed where the CFC is trying to minimise the distance between field and table. The CFC also engages in collective communication and decision-making around these activities; by doing so, it establishes linkages to political authorities, produces a project identity, and builds its members' political capital.

Notes

1 For Bang (2009, 123–126), this is a part of a bigger debate with the civic culture tradition represented first by Almond and Verba (1963). He criticises Almond and Verba (and others) for locating civic culture 'outside' and at the input side of the political system, whereas Bang himself (following Easton 1965) locates civic culture or political community *inside* the political system on a par and in an interplay with political authorities and the political regime.
2 In the text we refer to the interviews by a number, for example, interview person no. 1 is referred to as IP1, etc. The numbers following the IP reference refer to the relevant page in the interview transcript.
3 However, note that they carefully refrain from criticising traditional agriculture in their principles.
4 This is an attempt to create the oligopticon mentioned by Latour (2005, 175–183).
5 Unfortunately, this restaurant went bankrupt.
6 Bang (2009, 132) states that the credo of everyday makers is

> do it yourself, do it where you are, do it for fun, but also because you find it necessary, do it concretely instead of ideologically, do it self-confidently and show trust in yourself, do it with the system if need be.

This seems to express very neatly the content of the project identity of most CFC members. Interestingly, this interpretation is also confirmed by one of the founders of CFC, who has clearly educated himself in the scholarly literature on the subject:

> We have been what Bang and Sørensen (2003) call 'everyday makers', not political activists. We just try to solve a concrete problem in our everyday lives. This means that to the extent we feel part of a movement this is a really broad green movement.

(IP4, 10)

References

Allen, P. (2004) *Together at the Table: Sustainability and Sustenance in the American Agrifood System*. Philadelphia: Pennsylvania State University Press.

Almond, G. and Verba, S. (1963) *The Civic Culture.* London: Sage.

Bang, H.P. (2003) A new ruler makes a new citizen: cultural governance and everyday making. In H.P. Bang (ed.), *Governance as Social and Political Communication.* Manchester: Manchester University Press, 241–266.

Bang, H.P. (2009) 'Yes we can': identity politics and project politics for a late-modern world. *Urban Research & Practice*, 2(2), 117–137.

Bang, H.P. and Sørensen, E. (1999) The everyday maker: a new challenge to democratic governance. *Administrative Theory & Praxis*, 21(3), 325–341.

Bang, H.P. and Sørensen, E. (2001) The everyday maker: building social rather than political capital. In P. Dekker and E. Uslaner (eds), *Social Capital and Participation in Everyday Life.* London: Routledge, 148–161.

Beck, U. (1996) *The Reinvention of Politics.* Cambridge: Polity Press.

Beck, U. (2006) *Cosmopolitan Vision.* Cambridge: Polity Press.

Campbell, M.C. (2004) Building a common table. *Journal of Planning, Education and Research*, 23, 341–355.

Castells, M. (1997) *The Power of Identity.* Oxford: Blackwell.

Christensen, A.G. (2016) Landsbyfølelsen. Et casestudie af Københavns Fødevarefællesskab om betydningen af lokale fællesskaber for den senmoderne stedtilknytning. (The village feeling. A case study on the Copenhagen Food Coop and the importance of local communities for the late-modern sense of place). Master's thesis, University of Aalborg, Copenhagen.

De Vries, G. (2007) What is political in sub-politics? How Aristotle might help STS. *Social Studies of Science*, 37(5), 781–809.

DuPuis, M. and Goodman, D. (2005) Should we go 'home' to eat? Towards a reflexive politics in localism. *Journal of Rural Studies*, 21, 359–371.

Easton, D. (1965) *A Systems Analysis of Political Life.* Chicago: University of Chicago Press.

Hinrichs, C.C. (2003) The practice and politics of food system localization. *Journal of Rural Studies*, 19, 33–45.

Hoff, J. (2016) 'Think globally, act locally': climate change mitigation and citizen participation. In J. Hoff and Q. Gausset (eds), *Community Governance and Citizen-Driven Initiatives in Climate Change Mitigation.* London and New York: Routledge, 28–53.

Kenis, A. and Mathijs, E. (2014) (De)politicising the local: the case of the Transition Towns movement in Flanders (Belgium). *Journal of Rural Studies*, 34, 172–183.

King, C.A. (2008) Community resilience and contemporary agri-ecological systems: reconnecting people and food, and people with people. *Systems Research and Behavioral Science*, 25, 111–124.

La Cour, T. (2017) Et casestudie af Københavns Fødevarefællesskab. 'Strategisk branding og positionering med henblik på hvervning og fastholdelse af frivillige medlemmer'. Master's thesis, Department of Nordic Studies and Linguistic Sciences, University of Copenhagen, Denmark.

Latour, B. (2005) *Reassembling the Social – An Introduction to Actor-Network-Theory.* Oxford: Oxford University Press.

Latour, B. (2007) Turning around politics: a note on Gerard de Vries' Paper. *Social Studies of Science*, 37(5), 811–820.

Lloyd, A. and Pass, N. (2013) *Borgerlyst. Handlekraft i hverdagen.* Borgerlyst. Gylling: Narayana Press.

Marres, N. (2005) No issue, no public: democratic deficits after the displacement of politics. PhD Thesis, University of Amsterdam, the Netherlands.

Mol, A. (1999) Ontological politics. A word and some questions. *The Sociological Review*, 47(1), 74–89.

Mouffe, C. (2005a) *On the Political*. London: Routledge.

Mouffe, C. (2005b) *The Return of the Political*. London: Verso.

North, P. (2010) Eco-localisation as a progressive response to peak oil and climate change: a sympathetic critique. *Geoforum*, 41, 585–594.

Papazu, I. (2016) Participatory innovation: storying the renewable energy island Samsø. PhD Thesis, Department of Political Science, University of Copenhagen, Denmark.

Patton, M.Q. (1990) *Qualitative Evaluation and Research Methods*, 2nd edn. Newbury Park, CA: Sage.

Saunders, C. (2013) *Environmental Networks and Social Movement Theory*. London: Bloomsbury Academic.

Stoker, G. (2006) *Why Politics Matters*. Basingstoke: Palgrave Macmillan.

Swyngedouw, E. (2007) Impossible 'sustainability' and the postpolitical condition. In R. Krueger and D. Gibbs (eds), *The Sustainable Development Paradox*. London: The Guilford Press.

Swyngedouw, E. (2010) Apocalypse forever? *Theory, Culture, Society*, 27, 213–232.

Swyngedouw, E. (2013) The non-political politics of climate change. *ACME: An International Journal for Critical Geographies*, 12(1), 1–8.

Tavella, E. and Papadopoulos, T. (2017) Applying OR to problem situations within community organisations: a case in a Danish non-profit, member-driven food cooperative. *European Journal of Operational Research*, 258, 726–742.

Urry, J. (2011) *Climate Change and Society*. Cambridge: Polity Press.

Zizek, S. (1999) *The Ticklish Subject. The Absent Centre of Political Ontology*. London: Verso.

6 Rethinking environmentalism in a 'ruined' world

Lessons from the permaculture movement

Laura Centemeri

Introduction

Concepts and metaphors are important. They help us to frame problems and to imagine possible solutions. When speaking about environmental crises, everybody is familiar with the notions of externalities that must be internalised or risks one should be prepared for. Externalities, risks, adaptation, mitigation, resilience: these concepts shape the predominant approach public actors take to environmental issues. Other social actors, however, frame these crises in terms of 'cost-shifting' (Kapp 1983), socio-ecological injustices (Martinez-Alier 2002), damages to be repaired, and torts to be redressed. These frames entail substantially different ways of envisioning the actions required to tackle environmental problems.

I advocate that the notion of 'ruination' (Stoler 2013) is one possible way of understanding or making sense of the current condition of global socio-ecological crisis. In particular, I take this metaphor as the starting point for a reflection on environmentalism and the variety of its expressions in our societies (Guha and Martinez-Alier 1997; Armiero and Sedrez 2014).

I argue that, when confronted with processes of ruination, environmental engagement is not limited to *protection from*, or *denunciation against* the exploitation of work and nature. It also takes the shape of ordinary activities of *taking care of* people and their environments, involving practices of socio-ecological *repairing* and *regenerating*. Engagement in taking care of human and non-human beings and their environments is a way of directly experiencing the relevance of alternative ways to define what is worthwhile and to forge value arguments that differ from the dominant ones in that they are sensitive to contexts.

In fact, beyond its more perceptible forms, ruination points to an underlying process of *erosion of value arguments diversity* generated by the imposition of non-negotiable standardised rules and the forced marginalisation of context-sensitive value logics in the organisation of all spheres of social life. This erosion has been accelerated by the progressive hegemony acquired in recent decades by the neo-liberal form of 'governing through objective objectives' (Thévenot 2015).

The case of the permaculture movement serves as an example of a form of environmental engagement based on practices of *ecological care* (Puig de la

Bellacasa 2017). Departing from the results of an ongoing research project on the diffusion of permaculture in Italy, I examine how practices of ecological care in permaculture initiatives give rise to a variety of *pericapitalist economies*.[1] My third step addresses the issue of the risk of 'recuperation' (Boltanski and Chiapello 1999) of ecological care as a new source of legitimacy for capitalist accumulation. Ecological care can be reduced to just another argument supporting a 'green capitalism' or an 'economy of enrichment' (Boltanski and Esquerre 2017).

One way of avoiding recuperation is to build networks of pericapitalist economies while simultaneously multiplying the connections across the diverse forms of environmental engagement: practices of care, protest, denunciation, public participation, and lobbying. To do this, *political ecotones* must be designed at different scales. They are intended as identifiable spaces in which actors expressing diverse forms of environmentalism can meet on the basis of shared concerns. These spaces are of crucial importance to promote an effective coordination of diverse forms of environmental activism. More fundamentally, political ecotones can help the emergence of a shared socio-technical imaginary. In order to support the transition towards an 'ecological open society' (Audier 2017), this socio-technical imaginary should be able to combine the aspiration to socio-ecological justice with the call for emancipation and the development of 'an ethos of a more explicit acknowledgment of human immersion in non-human natural systems' (Schlosberg and Coles 2016, 166).

Ruination and the rise of 'governing through objective objectives': explaining the erosion of the diversity of value arguments

What form does environmental engagement take in a 'ruined' world? I started to explore this issue in my PhD thesis on the social responses to the Seveso, Italy, chemical contamination disaster. In particular, I investigated how the inhabitants of the small Italian town of Seveso reacted to a dioxin contamination following the industrial accident of 10 July 1976 at the ICMESA chemical factory. This small plant was owned by the Swiss company Hoffmann-La Roche. The disaster made a part of the city temporarily uninhabitable, caused a serious public health crisis, and triggered a mobilisation that turned the town of Seveso into the centre of the social struggles that were shaking Italy at that time, which included the legalisation of abortion and the denunciation of the 'crimes of capitalism' (Centemeri 2011, 2015).

In this post-disaster situation, I identified the relevance for the understanding of the recovery process of three forms of environmental activism: engaging in order to *protect* the environment from further destruction, *denouncing* responsibilities for what happened, asking for *reparation* (conviction of the guilty, monetary compensations, ecological restoration, symbolic reparation), and *regenerating* socio-ecological dynamics through ordinary practices of taking care of people, things, places, animals, and plants.

In Seveso, the actors who expressed these diverse forms of post-disaster environmental activism did not find a way to create synergies. Tensions fuelled divisive conflicts. But the experience of the failure in finding a collaboration between diverse expressions of environmentalism brought a small local group of activists (environmentalists and feminists) together to try to combine, over time, a practice of denunciation of environmental damage with one of valorising the *direct action of taking care* of a specific environment.

These activists saw practices of care as a way to express a non-confrontational critique, provided that they are guided by transformative purposes. In their view, taking care of an environment meant primarily repairing and regenerating socio-ecological relations, where repairing should not be seen as resistance to change. In their understanding, this particular practice of care not only implied resuming pre-existing relations and regaining previous life conditions, but was also seen as an opportunity to regenerate socio-ecological dynamics through *learning anew how to inhabit* an environment (Centemeri 2011). In particular, the development of a specific reflexivity on what one considers valuable in relation to the environment, not in abstract terms but in everyday activities, was considered as fundamental to the process of relearning how to inhabit.

What seems original to me in this approach to environmentalism is the importance attributed to the development of a reflexivity on what one considers valuable in relation to the environment: the way in which people define and attribute value to beings, things, places, and activities in their everyday practices becomes a key dimension in the process of repairing and regenerating. More specifically, practices of care require that people reconnect with context-sensitive value logics not primarily related to instrumental utility. Socio-ecological interdependencies connecting the local to the global and the sensory perception of places and people experienced in ecological-based processes of human life 'becoming' (Ingold 2000) were both considered as relevant perspectives to orient value judgements according to, respectively, a systemic logic of value and an 'emplaced' (Pink 2009) logic of value.

The emplaced logic of value rests on the experience of 'growing with' and 'knowing along' (Barua 2016) other human and non-human beings in situated 'encounters' (Haraway 2008). In this case, what constitutes value is linked to the development of familiarity with and affection for certain beings and places (Breviglieri 2012) or to the quality of an environment in terms of its atmosphere (Thibaud 2011) and as a place where the excitement of discovering something new is experienced (Auray 2016).

The point of the matter here is that a variety of logics of value and forms of knowledge – or 'pragmatic regimes of engagement' (Thévenot 2001, 2006, 2007) – are combined in practices of taking care of an environment. Through their in-context practices of care, people are consequently constantly engaged in a collective and ongoing process of *inquiry* in the Deweyan sense, which is meant to guarantee the creation and the maintaining of a diversity of goods, from 'emplaced' goods to 'public goods'.

Practices of taking care of an environment thus become a form of politics by other means since they offer to everyone the opportunity to explore and to deliberate on what counts as valuable in a variety of 'communities of care' that one can become involved with. They produce socio-ecological change primarily by means of the direct experience of the relevance of the diversity of value logics and practices in the shaping of sustainable ecologies and the resulting elaboration of alternative 'value arguments'.

By value arguments, I intend a recurrent reason or set of reasons supporting a certain understanding of what should count as valuable in a given situation. Following Francis Chateauraynaud (2015), it can be said that the strength of an argument is not simply based on intellectual coherence: it has to create a disposition to act. This means that value arguments always have a connection with an experiential substrate of value practices, understood as those practices through which actors (individually or collectively) define what they consider valuable and act accordingly to attain and maintain the condition deemed worthy (Dussage *et al.* 2015).

Far from being confined to the experience of the local group of activists in Seveso, the shift from values conceived as 'a realm somehow added from outside to material facts' to the understanding that they are 'immanent to the relations and orientations among moving beings [...] in a world of becoming' is common to a variety of 'new materialist movements' that are considered by Schlosberg and Coles (2016, 168) as expression of a 'new environmentalism of everyday life':

> These movements seek to critique and replace the devitalising and unsustainable practices of the domination of non-human nature with practices and flows that recognise human beings as animals in embedded material relationships with ecosystems and the non-human realm. The focus is on forging alternative, co-creative, productive and sustainable institutions at the local and regional level that reconstruct our everyday interactions with the rest of the natural world.
>
> (Schlosberg and Coles 2016, 173)

According to these authors, these new materialist movements are related predominantly to food and energy issues and to practices of recycling, repairing, and making the sphere of what they call 'new domesticity'.

In my view, underlying this transformation of environmentalism is the acknowledgment that the environmental crisis is related not only to random catastrophic events and episodic externalities but to a more ordinary and structural condition of progressive socio-ecological degradation. This ordinary process of degradation that, more often than not, goes unnoticed is linked to the dominance of 'growthism' not only as a guiding principle of economic organisation but as a central socio-technical imaginary.

Rob Nixon (2011, 4) coined the notion of 'slow violence' to point to those processes that create the conditions of 'conjoint ecological and human disposability' or, to put it differently, of 'simplification for alienation' (Tsing 2015), that are necessary to sustain growth as currently intended, i.e. as measured by GDP. My point is that this disposability by means of simplification entails reducing the *legitimate value arguments* that one can resort to in public decision-making processes in order to justify certain value practices against others.

Justifiable value practices are those value practices that are socially encouraged, routinised, and stabilised through 'investments in forms' (Thévenot 1984), the establishment of conventions (Diaz-Bone and Salais 2011), and rules of socialisation so as to maintain the conditions for the reproduction of a certain socioeconomic order. They rest on shared legitimate value arguments and shared 'socio-technical imaginaries' (Jasanoff and Kim 2015): these arguments and imaginaries combine logics or 'modes of valuation' (Centemeri 2017) in specific ways that are privileged over others.

In the current phase of global capitalism, dominant value practices are influenced by arguments and imaginaries that establish the uncontested centrality of economic growth, which is recognised as the source of social well-being. As a consequence, social goals are translated into standardised and quantified objectives to be achieved *under budgetary constraints*, giving rise to what Laurent Thévenot (2015) calls 'governing through objective objectives'. The process whereby good government is defined as the achievement of 'objective objectives' accounts for the progressive erosion of the diversity of legitimate value arguments in the public space and for public decision-making being understood as the result of calculation and not deliberation. As discussed by Torre (2018), these processes entail the steady transformation of sovereignty into a generalised colonial model in which increasing social inequalities are the norm (see Piketty 2014).

The phenomenon of ruination is directly connected with the reduction of the spaces of deliberation on legitimate value arguments, since this de facto erosion of the diversity of legitimate value arguments has an impact on value practices and, consequently, on the shaping of socio-ecological systems. Ruined socio-ecological systems have lost their diversity (in terms of populations, functions, and interdependencies): they are off-balance, they have been exposed to contaminations, and they are at risk of breakdown. Their ruination, however, has provided the conditions for profitable exchange and economic growth (Tsing 2015).

This change in the modes of governing helps explain why value practices orienting ordinary activities have now become a crucial concern in many environmental movements. In particular, these movements identify the promotion of practices of ecological care as a possible way to deal with what remains in place, notwithstanding ruination, in such a way as to trigger processes of repairing, regenerating and, potentially, resisting.

Environmentalism in the face of ruination: understanding the perspective of care

The notion of *care* is increasingly used in environmental discourses to frame the understanding of the relationship between societies and their environments. My hypothesis is that, together with risk and limit, the notion of care expresses a specific form of 'environmental reflexivity' (Charbonnier 2017). According to French philosopher Pierre Charbonnier, 'risk' and 'limit' are the two central notions that have been mobilised in social sciences to frame the relationship between societies and their environments and to understand the ecological crisis. Today, he contends, social sciences find in the concept of 'Anthropocene' (Steffen *et al.* 2011) a new frame with which to understand the ecological crisis. Anthropocene, originally a geological notion, means that humanity is now considered as a geophysical force causing climate change, massive erosion of biodiversity, and the depletion of natural resources.

However, as emphasised by French sociologists Francis Chateauraynaud and Josquin Debaz (2017, 585), the adoption of the paradigm of the Anthropocene can engender at least three possible epistemic and axiological attitudes:

- an attitude of control of the 'Earth system', which is in line with the 'risk paradigm' and its technocratic drift;
- an apocalyptic attitude connected with the paradigm of limits and the perspective of collapse; and
- an attitude that pays attention to the irreducible variety of interdependencies that bind human beings to their environments and shape the latter as places to live together with other human and non-human beings.

This last approach, which the authors provocatively term 'counter-Anthropocene', consists of paying attention to the observable plurality of ways in which human beings create a variety of 'micro-worlds' (Chateauraynaud and Debaz 2017, 601), that is, socio-ecological systems not completely determined by existing *dispositifs* but expressing a capacity for 'self-government' (Zask 2010).

Practices of care are fundamental to the emergence of micro-worlds in which human needs can be met while guaranteeing the conditions in which other species (animal, vegetable) can also thrive. In fact, according to the definition of the feminist political theorist Joan Tronto, caring is:

> a species activity that includes everything that we do to maintain, continue, and repair our world so that we can live in it as well as possible. That world includes our bodies, our selves, and our environment, all of which we seek to interweave in a complex, life-sustaining web.
>
> (Tronto 1993, 103)

Caring thus implies attention, concern, solicitude. Following Tronto, as *caring about* the other (human or non-human), care is a way of perceiving the world that consists of paying attention to what the other needs. As *taking care*, it is a way of being concerned about others, which implies assuming responsibility. As *care giving*, it is a way of concretely taking care of the other, which implies the exercise of a competence.

In contrast to the primacy attributed to the ideal of autonomy in classical political and moral thinking, the care approach sees the individual as the result of multiple interdependencies, not only with other human beings but also with the environment. This ecological vision of the human being leads to the recognition of a condition of vulnerability of the human life form and the socio-ecological systems sustaining it.

The concept of 'ecological care', introduced by Maria Puig de la Bellacasa (2017), synthesises the type of practical engagement with the environment that is required to maintain the complex network of interdependencies that support life on earth, starting from the life of soils. She argues that if we take life dynamics, or 'bios', seriously as a matter of concern, then the traditional notion of collective is challenged and it has to include other than human beings.

As a mode of engaging with the environment, care calls for the daily exercise of concerned attention, or awareness of the vulnerability that is peculiar to the human form of life, which is of a relational and ecological nature. This attention becomes an action of support where necessary. This action must be guided by the comprehension of the specificities of the context rather than by abstract rules. Care is therefore always contextual and not essentialist.

Moreover, once applied to the environment, it becomes clear that care does not mean generalised love or compassion, it also means choice, exclusion, and struggle (Tsing 2012a). Ecological care requires coping with plagues as well as the need to ally with certain beings against others. Care is not immune to tragic choices; on the contrary, it implies an increased awareness of the tragic dimension of life, as discussed by philosopher Martha Nussbaum (2001).

According to Tronto (2012, 9), care can 'free us from incessant refrains about our powerlessness to act' in the face of the socio-ecological crisis. However, practices of care can be articulated in conservative or emancipatory political endeavours. It depends on the social imaginaries that connect alternative value practices of care with a larger normative horizon shared with others.

For example, the encyclical letter that Pope Francis recently devoted to the ecological crisis is entitled *Laudato Sii. On care for our common home*. In the case of this encyclical, which has very strong anti-capitalist tones, the perspective of ecological care is translated into the idea of an 'integral ecology' to point to the bond existing between humans and the natural world and the need for an integrated approach to environmental problems and social justice issues.

Not surprisingly, however, and in line with the Catholic doctrine, a central dimension of this integral ecology is the complementary union of man and woman in heterosexual marriage, seen as the expression of the natural order of things. In the political vision delivered by the Pope, practices of taking care of the environment are, therefore, combined with an imaginary in which, at some point, there is a natural order to be respected and protected. Depending on the kind of collective actors that mobilise the encyclical letter in the political arena, some may stress the need to promote socio-ecological justice while others emphasise the conservative tone of the integral ecology perspective. Although both can express a critique of capitalism through care as the main value argument, they may be on opposite sides in terms of emancipatory struggles.

In particular, the emancipatory understanding of practices of care is based on the idea that engaging with the environment through ecological care produces 'naturecultures',[2] an approach that comes from an (eco)feminist tradition. In this case, there is no such thing as a 'natural order' to respect but a co-construction of a variety of local naturecultural orders that are always in the making. The imaginary of naturecultures, however, is not per se emancipatory. As Luigi Pellizzoni notes, the contingency and indeterminacy that are implicit in this vision 'resonates with the way in which science and the biophysical world are being "neoliberalized"' (Pellizzoni 2014, 83, see also Pellizzoni 2015).

For a better understanding of how this ecological care perspective can inspire and influence environmental engagement in practice and make it emancipatory, I decided to combine this more theoretical exploration about care and the environment with research on the permaculture movement, for which the three guiding ethical principles for an ecological way to live are: *earth care, people care,* and *fair share.* As a result, I started a research project in 2015 on the permaculture movement as an environmental and transnational movement based on practices of ecological care. More specifically, I am currently doing research on the diffusion of permaculture in Italy.

My analysis of permaculture initiatives is based on several data sources as well as a large corpus of permaculture books and writings (in English, French, and Italian). Data were collected through semi-structured interviews I conducted with permaculture activists (in Italy, France, Spain, Portugal and Australia); a survey administered to Italian permaculturists in December 2016; participation in national and international permaculture 'convergences' (that is, regular meetings of permaculture associations); attendance at a Permaculture Design Course in Catania and of two other permaculture courses in Milan (Italy); and the direct observation of permaculture initiatives in Italy. Other data were collected through the analysis of permaculture activists' blogs, permaculture groups' social media pages (especially Facebook) and permaculture magazines and webzines devoted to transition issues.

Reinhabiting: promises and perils of permaculture's alternative value practices

Permaculture – a term that derives from the contraction of 'culture' and 'permanent' – is a concept that originated in Australia, specifically in Tasmania, and was developed in the 1970s by Bill Mollison (1928–2016), an eclectic environmental psychology professor, and David Holmgren (1955), Mollison's student at Hobart University (Mollison and Holmgren 1978; Mollison 1988; Holmgren 2002).

The 1970s in Australia, as elsewhere, were years of environmental struggles and counter-cultural movements, including the highly composite 'back to the land' movement (Calvário and Otero 2015). Permaculture was first conceived as a support tool to facilitate these 'returns to the land' so that people with very little familiarity with agricultural practice could settle in rural areas and develop subsistence economies based on agriculture.

But permaculture is not a set of techniques for agriculture. It is a *method* to *design the organisation of basic human activities* (food, health, education, housing, agriculture, forestry, etc.) in such a way that they are not simply sustainable but perennial. In permaculture, human activities are seen as generating socio-ecological systems that should be designed in a permacultural way. A permacultural design aims to reduce and optimise the need for energy input (including work), increase diversity (of functions, populations, species, etc.) and resilience, and to guarantee an abundance of diverse and varied forms of both material and immaterial wealth. This is done through imitating problem-solving strategies that are observable, or have been observed, in 'healthy' ecosystems. This is a principle usually defined as of *biomimicry* even if 'ecomimicry' would be a more accurate description of such design practices.

In Mollison's and Holmgren's view, the 'back to the landers' of the 1970s needed, above all, frameworks to guide the 'reintegration' of their activities in ecosystems in such a way as to be able to trigger virtuous processes of coevolution. This reintegration in ecosystems is basically a change in the way people individually and collectively respond to basic needs and, more generally, how they organise human activities. This is not just a matter of techniques but also of what I have previously defined as value practices.

Permaculture value practices are based on the combination of *ethical principles*, *design principles*, and *attitudinal principles*. Design principles are rules of thumb and problem-solving strategies inspired by the experience of human communities' management of environmental resources all over the world and the observation of healthy ecosystems. David Holmgren (2002) provides a concise list of 12 basic permaculture design principles, including 'use and value diversity', 'use edges and value the marginal', and 'creatively use and respond to change'. They partially overlap with attitudinal principles that aim to develop a practical wisdom in the approach to the design of complex socio-ecological systems, such as 'work with nature, not against it' and 'the problem is the solution'.

Design and attitudinal principles help define strategies of action that must be guided by principles of 'earth care', 'people care', and 'fair share' (or return of the surplus) – the three ethics of permaculture. Earth care, people care, and faire share are the main value arguments in permaculture. These value arguments of *care* and *distributive justice* are extended from humankind to the many non-human beings and entities that guarantee life is maintained in ecosystems, since 'all are our family' (Mollison 1988, 3). Consequently, earth care in permaculture means the care of soil, which is in turn understood as a heterogeneous living community. I propose the term 'multispecies commoning' (Centemeri 2018) to describe the practices of mutualist and non-antagonistic interspecies entanglements that permaculture promotes through its design method and ethics. This implies the adoption of appropriate value practices.

My observations confirm that an important part of permaculture teaching and training involves an investment in reawakening the attention given to the importance of what and how we value, with the goal of helping to recognise the diversity of modes of valuation and value arguments that should inform a permacultural design. In permaculture teaching and training, the logics of valuation are taken beyond human-centred universal or standardised goal-oriented understandings of value to stress the importance of context-sensitive modes of valuation, including what I have previously defined as 'emplaced modes of valuation'.

In permaculture design, it is essential to nurture plural perspectives on modes of valuation, from the most universal to emplaced modes. Therefore, a crucial skill in permaculture is the capacity to combine diverse logics of value in the organisation of human activities with the aim of renewing the bonds of positive collaborations with the 'biotic community' (Leopold 1949). At the same time, permaculturists also strive to avoid 'remoteness'. Remoteness, as discussed by ecofeminist theorist Val Plumwood (2002, 77), is the belief that given the right conditions, 'living close to the land' is the way to 'generate knowledge of and concern for ecological effects of production and consumption within a local community'. However, as Plumwood emphasises:

> neither this closeness nor the local ecological literacy it might help generate is sufficient to guarantee knowledge of ecological effects and relationships in the larger global community or even a larger regional one. This requires a larger network whose formation seems unlikely to be assisted by economic autarchy.
>
> (Plumwood 2002, 77)

Permaculture is not simply a technique to regenerate soil. It is conceived and transmitted through teaching, training and 'demonstration' (Rosental 2013) as an *art* (de Certeau 1990) meant to repair socio-ecological systems and to help people *reinhabit* them. At the same time, it is a movement that in order to avoid remoteness tries to foster the creation of networks of reinhabitants and progressively larger 'networks of networks' through which alternative institutions can emerge.

'Reinhabitation' is a concept drawn from the intellectual tradition of American bioregionalism. According to anthropologists Joshua Lockyer and James R. Veteto, reinhabitation entails:

> a process whereby individuals and communities decide to commit themselves to a particular bioregion and live 'as if' their descendants will be living there thousands of years into the future. [...] Bioregionalists often take the indigenous societies of their bioregions as models of long-term inhabitation and sustainability, but work within their own cultural traditions, with a sense of dynamism that does not reify or essentialise traditional place-based cultures.
>
> (Lockyer and Veteto 2013, 8–9)

Through the concept of reinhabiting, Lockyer and Veteto underline the fact that, together with other 'ecotopian' and new materialist movements, the permaculture movement promotes an understanding of both politics and the economy from 'the standpoint of place' but at the same time has an emancipatory vision: 'anyone of any race, any religion, or origin is welcome, as long as they live well on the land…. This sort of future culture is available to whoever makes the choice, regardless of background' (Lockyer and Veteto 2013, 34).

Choosing to be committed to a place (and live well on it) entails being engaged in creating alternative social and economic institutions and nurturing networks. These institutions and networks ensure certain forms of circulation between 'localities' to share knowledge, goods, and experience, and support struggles for global justice. They are a reaction against the generalised marketisation of all aspects of life induced by dominant value practices. The creation of alternative institutions and networks also represents a challenge to the politically reactionary anti-modernist understanding of place as the 'homeland' determining the entirety of one's identity, past, present, and future in an immutable natural order of things.

Observation of Italian permaculture initiatives confirms that the *art of reinhabiting* requires the development of alternative value practices that rest primarily on context-sensitive modes of valuation. This implies that they are 'nonscalable', that is, they are permeable to the diversity of contexts and the indeterminacies that originate from the encounter with this diversity (Tsing 2012b). They challenge locally the hegemony of current dominant value practices but they are always at risk of being recuperated by capitalism.

Like the autonomous urban social centre activists interviewed by Paul Chatterton (2010, 1216), Italian permaculturists also share the same condition of 'dwell[ing] both in the hoped-for and actual world': this implies paying attention to a 'more complex and subtle understanding of anti-capitalist practice'. 'Post-capitalist' and 'despite-capitalist' initiatives should also be taken into account as an expression of a non-confrontational critique.

In line with this more complex and subtle understanding of anti-capitalism, my fieldwork shows how Italian permaculture initiatives leads to the emergence of local 'pericapitalist' (Tsing 2015) economies,[3] in which alternative value practices are combined with dominant value practices. Monetary returns are important but not as a goal per se. There are frequent discussions concerning the 'just price', especially for the tuition fees to attend permaculture trainings. Diverse solutions are provided to the problem of just price with no shared guidelines, with the exception of a generic call for transparency and fairness. Those permaculturists that are actively involved in the movement usually justify their prices, 'accept feedbacks' (a permaculture principle) and discuss collectively about it. They share the concern that permaculture risks becoming just another niche in the market of consultancy and professional training, generating mechanisms of competition for profit. Even if some of them consider the market as a potential ally to spread permaculture (according to the permaculture principle that 'the problem is the solution'), there is general agreement on the need to reduce consumption and limit profit accumulation.

More generally, the place of dominant value practices in permaculture initiatives seems to become less important when they are integrated into networks of pericapitalist economies that support cooperation and 'mutual aid' across localities. Equally important is the participation in networks that openly try to challenge dominant value practices and arguments through also resorting to protest. The *Genuino Clandestino* network (De Angelis 2017, 294–300) is one such example. Originating from the 2001 initiative of activists in the Italian city of Bologna concerned about the issue of 'food sovereignty', this network challenges official systems of food certification, especially for processed foods, through the creation of participatory systems of self-certification and a network of farmer markets distributing self-certified products. It constitutes an alternative economic institution intended to support local initiatives of regenerative agriculture, whether or not they are inspired by permaculture.

Actors in the *Genuino Clandestino* network try to develop not just alternative economies but *subversive economies*, in that producers and consumers are considered as 'commoners'. This means the value practices that structure the socioecological system they become part of are not oriented towards producing profit but a diversity of ecological and social forms of (common)wealth from which all participants can benefit.

Nevertheless, some of the permaculturists I have met express a certain pessimism about being able to change political and economic actors' visions of value so that other measures of value besides profit are given importance. They believe that the ability to produce economic results is the best proof of the social desirability of permaculture, especially in agriculture. Although these permaculturists seem quite disenchanted, particularly in relation to professional farmers, they try to find opportunities to engage with this public regardless of any initial hostility.

For example, Gautier is a French permaculturist living in the Catania province in Sicily who is trying to develop alternative fertilisers based on fish fermentation. He not only promotes them by showing traditional farmers that they are less expensive and more effective but also provides a concrete example of the principles of the circular economy through offering a solution to dispose of spoiled fish. Gautier believes that being able to influence standard production practices (such as fertilising practices) at the margins is an important step towards a wider change, even though value practices are not directly addressed. In his view, promoting agricultural regenerative practices is considered a realistic enough objective.

However, this emphasis on the good of ecological regeneration per se risks eclipsing the importance of issues of 'fair share' and 'people care'. Permaculturists should be concerned in their initiatives also about issues of social inequalities and forms of exploitation other than soil depletion.

These concerns have a central place in the *Saja* permaculture project started by Salvo and a group of friends in their 30s in Paternò (Sicily) in 2011. Together, they have gradually turned an abandoned citrus grove of 1.8 hectares that was bought by Salvo into a diversified polyculture. In Salvo's own words, this citrus grove – called in Sicilian dialect '*u Jardinu*' (the garden) – is 'a place of production and contemplation' and 'an oasis of diversity, sociability and resilience' where agriculture is practised according to principles of collaboration and cooperation with human and non-human beings. The *Saja* project also promotes a culture of hospitality, of sharing knowledge, and mutualising competencies and resources while valorising diversity. In Salvo's vision, this different approach to agricultural work at *Saja* should contribute not only to ecological regeneration but to changing the imaginary of agriculture in the local community, especially among the young. In fact, oranges are the dominant production in Paternò through an intensive monocultural agriculture under very poor working conditions. Through promoting a local network of alternative agricultural projects that adopt the same culture of hospitality, sharing knowledge, mutualising competencies and resources, and 'multispecies commoning', Salvo aims to contribute to the local production of broader cultural change by demonstrating the virtues of ecological care.

The alternative value practices that I observed in Italian permaculture initiatives are articulated with alternative value arguments and emerging sociotechnical imaginaries. Recurrent topics in these arguments and imaginaries are 'transition', 'degrowth', 'slow living', 'conviviality', 'living in harmony with nature', 'abundance', 'rurbanity, inclusion and emancipation, and also 'collapse', self-sufficiency, and natural order.

There is a certain syncretism in these arguments and imaginaries. This accounts for the fact that some permaculture initiatives focus mainly on self-sufficiency while other permaculture-inspired projects evolve in niches of the market economy. Whereas some actors believe that you can only change the system by stepping out of it, others think that you have to transform the system from within or build alternative institutions. These approaches to social change coexist in the Italian permaculture community, but not without friction. They pertain to diverse, and sometimes conflicting, *political cultures*. Political cultures can be defined as shared social imaginaries and common 'styles of action that organize political claims-making and opinion-forming, by individuals or collectivities' (Lichterman and Cefaï 2006). In permaculture terms, political cultures can be considered 'invisible structures', that is, as part of the cultural elements that influence how we relate to other people and the environment. However, they are not explicitly addressed in permaculture thinking either as obstacles or resources for designing effective collective action for change.

Networks of networks and political ecotones against the risks of recuperation

Permaculture is generally considered as the expression of a form of environmentalism in which direct actions of repairing, caring, and reinhabiting are privileged over more classical forms of political engagement such as protest or lobbying. Instead of being focused on denouncing what is wrong in the current situation, permaculturists are supposed to be positively engaged in taking responsibility and changing things through creating new forms of organisation of human activities that will hopefully make the existing ones obsolete. Similar to other neo-materialist movements, in permaculture initiatives 'the development of community movements and institutions – beyond solely individualized action – is purposeful and pointed' (Schlosberg and Coles 2016, 165).

What emerges from my Italian fieldwork is that some permaculture practitioners consider individual direct action as the only meaningful way to be politically engaged. These permaculturists show a profound distrust in collective action, including in the permaculture movement and in public institutions. This can lead to a form of 'socially conservative individualism', similar to what Matthew Schneider-Mayerson (2015) observed in the population of US 'peakists'. These are usually 'do it yourself' permaculturists who do not participate actively in the associative life of the permaculture movement and refuse formal permaculture training.

In other cases, permacultural practices of reinhabiting are considered as complementary to other more classical forms of environmental engagement. They make it possible to reach people who are not into environmentalism through practical activities and the demonstration of the existence of alternative ways of organising subsistence activities. However, in order to promote social change towards sustainability, policies have to be modified and governments must be convinced to sustain grassroots initiatives. This implies challenging current patterns of power organisation and wealth distribution. More traditional forms of collective action, such as lobbying and protesting, are required.

To reach these objectives, the permaculture movement supports a strategy of collaboration, coalition, and alliances with other collective actors, providing tools and methodologies to organise 'networks of networks', designing them as ecosystems whose resilience is related to diversity, and the capacity to recognise and sustain emergent positive 'patterns'. This is done through openness to a certain level of 'hybridisation' (Tosi and Vitale 2009) and the active involvement of permaculturists in other political movements, public institutions, international organisations and in the private sector.

Equally important is the development of what I call 'political ecotones', that is, physical spaces – such as collective gardens, recurrent gatherings such as festivals or fairs, certain local markets, and occupied urban open spaces – that can be analysed as transition areas between diverse forms of environmental engagement and diverse political cultures. Through bringing together a variety of people in a specific place around ecological care practices – such as agroecology, bioconstruction, or local food networks – the idea is to get citizens and activists of diverse political cultures to get to know each other and to recognise common concerns.

In these political ecotones and 'networks of networks' that are designed to encourage the expression of diversity, shared value arguments and socio-technical imaginaries start to take shape and try to gain critical strength. But conflicts also emerge, which demonstrates the difficulty of finding a shared normative horizon.

This is a controversial topic in the permaculture movement. In fact, permaculture founders supported a vision of a non-polarised and non-contentious politics based on the assumption that 'it is possible to agree with most people, of any race or creed, on the basics of life-centered ethics and commonsense procedures, across all cultural groups' (Mollison 1988, 508). This 'post-political' (Swyngedouw 2010) belief in the power of life to create a spontaneous alignment underestimates the fact that 'life-centred ethics' can be reactionary and not necessarily emancipatory. How should we deal with governments that deny fundamental democratic values while promoting ecological regenerative practices?

In this respect, a crucial issue seems to be the way in which care and its relation to nature is conceived. In permaculture earth care, nature is presented as both a teacher and a partner. References to nature seem to oscillate between an understanding of nature as an expression of an order we have to respect or as an 'assemblage' (Dodier and Stavrianakis 2018), a composite web of life, a variety of 'naturecultures' emerging from practices of ecological care. These two opposite understandings of nature – as order and as naturecultural assemblage – entail diverse interpretations of the place of care in the construction of the political community as a politically conservative or a potentially emancipatory value argument. But the emancipatory potential of care as value argument can express itself only if care is combined with fair share, that is, with the taking into account of solidarities and issues of social and ecological justice.

Permaculture is committed to transforming the modern understanding of the relationship between society and nature, which goes hand in hand with the modern understanding of the relationship between knowledge and experience and between reason and emotion. Permaculture thinking tries to rethink these dichotomies, not as static oppositions but as dynamic and moving frontiers (Cohen 2018). It is necessary to make this radical challenge to some of the fundamental pillars of Western modern culture while preserving a commitment to social justice and individual freedoms so that a new social imaginary that is both ecological and emancipatory can emerge.

French philosopher Serge Audier (2017) discusses the notion of 'ecological open society' to define a society in which a social imaginary of justice and emancipation is articulated with the taking into account of ecological interdependencies. More precisely, from my research on the permaculture movement, the real stake appears to be that of finding a way to combine the aspiration to socio-ecological justice with the call for emancipation and the development of an ethos that acknowledges that we as humans are immersed in non-human natural systems (Schlosberg and Coles 2016, 166).

The complexity of the cultural challenge raised by permaculture mirrors the severity of the crisis we are living through, in which ecological, financial and social dimensions are interwoven (Fraser 2014). The response to this crisis requires change to the personal and the political. It requires reinventing our value practices and our ways of conceiving what worth is while re-imagining the political community as the result of practices of 'multispecies commoning' (Centemeri 2018) grounded in the aspiration to social justice and emancipation. Even if local pericapitalist economies structured by alternative value practices and arguments do not represent the alternative, they 'can be sites for rethinking the unquestioned authority of capitalism in our lives. At the very least, diversity offers a chance for multiple ways forward—not just one' (Tsing 2015, 65). However, an economic system that recognises the legitimacy of non-scalable value practices and context-sensitive value arguments is, without doubt, not easily compatible with the current form of capitalism.

Notes

1 This contribution presents some of the results of the research programme SYMBIOS – Social Movements For The Transition Towards A Frugal Society, directed by Gildas Renou (University of Strasbourg) and funded by the French ANR (ANR-14-CE03–0005–01). Additional funding was provided by the Ecole Française de Rome and the CNRS international mobility programme. I wish to thank Gildas Renou and Anders Blok for their comments on a previous version of this paper.
2 'Natureculture' is a term that points to the inseparability of the natural and the cultural against an ontological split largely supposed in modern traditions (Haraway 1991; see also Latour 1993).
3 Here, 'economy' is understood as an institutionalised interaction between human beings and their natural surroundings, providing them with the conditions to satisfy material wants (see Polanyi 1977).

References

Armiero, M. and Sedrez, L. (eds) (2014) *A History of Environmentalism: Local Struggles, Global Histories.* New York: Bloomsbury.

Audier, S. (2017) *La société écologique et ses ennemies. Pour une histoire alternative de l'émancipation.* Paris: La Découverte.

Auray, N. (2016) *L'alerte ou l'enquête.* Paris: Presses des Mines.

Barua, M. (2016) Lively commodities and encounter value. *Environment and Planning D,* 34(4), 725–744.

Boltanski, L., and Chiapello, E. (1999) *Le nouvel esprit du capitalisme.* Paris: Gallimard.

Boltanski L. and Esquerre, A. (2017) *Enrichissement. Une critique de la marchandise.* Paris: Gallimard.

Breviglieri, M. (2012) L'espace habité que réclame l'assurance intime de pouvoir. *Études Ricoeuriennes/Ricoeur Studies,* 3(1), 34–52.

Calvário, R. and Otero, I. (2015) Back-to-the-landers. In G. D'Alisa, F. Demaria and G. Kallis (eds), *Degrowth: A Vocabulary for a New Era.* Oxford and New York: Routledge.

Centemeri, L. (2011) Retour à Seveso. La complexité morale et politique du dommage à l'environnement. *Annales. Histoire, Sciences Sociales,* 66(1), 213–240.

Centemeri, L. (2015) Investigating the 'Discrete Memory' of the Seveso Disaster in Italy. In S. Revet and J. Langumier (eds), *Governing Disasters. Beyond Risk Culture.* London: Palgrave Macmillan.

Centemeri, L. (2017) From public participation to place-based resistance. Environmental critique and modes of valuation in the struggles against the expansion of the Malpensa airport. *Historical Social Research,* 42(3), 97–122.

Centemeri, L. (2018) Commons and the new environmentalism of everyday life. Alternative value practices and multispecies commoning in the permaculture movement. *Rassegna Italiana di Sociologia*, 64(2), 289–313.

Charbonnier, P. (2017) Généalogie de l'Anthropocène. La fin du risque et des limites. *Annales. Histoire, Sciences Sociales*, 72(2), 301–328.

Chateauraynaud, F. (2015) Environmental issues between regulation and conflict. Pragmatic views on ecological controversies. *GSPR Working Paper Series*, EHESS.

Chateauraynaud, F. and Debaz, J. (2017) *Aux bords de l'irréversible. Sociologie pragmatique des transformations*. Paris: Petra.

Chatterton, P. (2010) So what does it mean to be anti-capitalist? Conversations with activists from urban social centres. *Urban Studies*, 47(6), 1205–1224.

Cohen, A. (2018) Usage des oxymores et pratique des lisières. *Cahiers philosophiques*, 2(153), 25–37.

De Angelis, M. (2017) *Omnia Sunt Communia On the Commons and the Transformation to Postcapitalism*. London: Zed Books.

de Certeau, M. (1990) *L'invention du quotidien*. Paris: Gallimard.

Diaz-Bone, R. and Salais, R. (2011) Economics of convention and the history of economies. Towards a transdisciplinary approach in economic history. *Historical Social Research*, 36(4), 7–39.

Dodier, N. and Stavrianakis, A. (eds) (2018) *Les objets composes. Agencements, dispositifs, assemblages*. Paris: Editions de l'EHESS.

Dussage, I., Helgesson, C., and Lee, F. (eds) (2015) *Value Practices in the Life Sciences and Medicine*. Oxford: Oxford University Press.

Fraser, N. (2014) Can society be commodities all the way down? Post-Polanyian reflections on capitalist crisis. *Economy and Society*, 43(4), 541–558.

Guha, R. and Martínez-Alier, J. (1997) *Varieties of Environmentalism: Essays North and South*. London: Earthscan.

Haraway, D.J. (1991) *Simians, Cyborgs, and Women: The Reinvention of Nature*. New York: Routledge.

Haraway, D.J. (2008) *When Species Meet*. Minneapolis, MN: University of Minnesota Press.

Holmgren, D. (2002) *Permaculture. Principles and Pathways beyond Sustainability*. Hepburn: Holmgren Design Services.

Ingold, T. (2000) *The Perception of the Environment: Essays on Livelihood, Dwelling and Skill*. London: Routledge.

Jasanoff, S. and Kim, S.-H. (eds) (2015) *Dreamscapes of Modernity: Sociotechnical Imaginaries and the Fabrication of Power*. Chicago: University of Chicago Press.

Kapp, K.W. (1983) *Social Costs, Economic Development and Environmental Disruption*. Lanham: University Press of America.

Latour, B. (1993) *We Have Never Been Modern*. Cambridge, MA: Harvard University Press.

Leopold, A. (1949) *A Sand County Almanac*. New York: Oxford University Press.

Lichterman, P. and Cefaï, D. (2006) The idea of political culture. In R.E. Goodin and C. Tilly (eds), *The Oxford Handbook of Contextual Political Analysis*. Oxford: Oxford University Press.

Lockyer, J. and Veteto, J.R. (eds) (2013) *Environmental Anthropology Engaging Ecotopia. Bioregionalism, Permaculture, and Ecovillagies*. New York and Oxford: Berghahn.

Martinez-Alier, J. (2002) *The Environmentalism of the Poor. A Study of Ecological Conflicts and Valuation*. Cheltenham: Edward Elgar.

Mollison, B. (1988) *Permaculture – A Designer's Manual*. Tyalgum: Tagari Publications.

Mollison, B. and Holmgren, D. (1978) *Permaculture One: A Perennial Agricultural System for Human Settlements*. Tyalgum: Tagari Publications.

Nixon, R. (2011) *Slow Violence and the Environmentalism of the Poor*. Cambridge and Oxford: Oxford University Press.

Nussbaum, M. (2001) *The Fragility of Goodness: Luck and Ethics in Greek Tragedy and Philosophy*. Cambridge: Cambridge University Press.

Pellizzoni, L. (2014) Metaphors and problematizations notes for a research programme on new materialism. *Tecnoscienza. Italian Journal of Science & Technology Studies*, 5(2), 73–91.

Pellizzoni, L. (2015) *Ontological Politics in a Disposable World. The New Mastery of Nature*. Farnham: Ashgate.

Piketty, T. (2014) *Capital in the Twenty-first Century*. Cambridge, MA, and London: Harvard University Press.

Pink, S. (2009) *Doing Sensory Ethnography*. Thousand Oaks, CA: Sage.

Plumwood, V. (2002) *Environmental Culture: The Ecological Crisis of Reason*. London: Routledge.

Polanyi, K. (1977) *The Livelihood of Man*. New York: Academic Press.

Puig de la Bellacasa, M. (2017) *Matters of Care: Speculative Ethics in More than Human Worlds*. Minneapolis: University of Minnesota Press.

Rosental, C. (2013) Toward a sociology of public demonstrations. *Sociological Theory*, 31(4), 343–365.

Schlosberg, D. and Coles, R. (2016) The new environmentalism of everyday life: sustainability, material flows, and movements. *Contemporary Political Theory*, 15(2), 160–181.

Schneider-Mayerson, M. (2015) *Peak Oil. Apocalyptic Environmentalism and Libertarian Political Culture*. Chicago and London: The University of Chicago Press.

Steffen, W., Grinevald, J., Crutzen, P., and McNeill, J. (2011) The Anthropocene: conceptual and historical perspectives. *Philosophical Transactions of the Royal Society A*, 369(1938), 842–867.

Stoler, A.L. (ed.) (2013) *Imperial Debris. On Ruins and Ruination*. Durham and London: Duke University Press.

Swyngedouw, E. (2010) Apocalypse forever? Post-political populism and the spectre of climate change. *Theory, Culture & Society*, 27(2–3), 213–232.

Thévenot, L. (1984) Rules and implements: investment in forms. *Social Science Information*, 23(1), 1–45.

Thévenot, L. (2001) Pragmatic regimes governing the engagement with the world. In K. Knorr-Cetina, T. Schatzki and E.V. Savigny (eds), *The Practice Turn in Contemporary Theory*. London: Routledge.

Thévenot, L. (2006) *L'action au pluriel. Sociologie des régimes d'engagement*. Paris: La Découverte.

Thévenot, L. (2007) The plurality of cognitive formats and engagements: moving between the familiar and the public. *European Journal of Social Theory*, 10(3), 413–427.

Thévenot, L. (2015) Certifying the world: power infrastructures and practices in economies of conventional forms. In P. Aspers and N. Dodd (eds), *Re-Imagining Economic Sociology*. Oxford: Oxford University Press.

Thibaud, J.-P. (2011) A sonic paradigm of urban ambiances. *Journal of Sonic Studies*, 1(1). Available at: http://journal.sonicstudies.org/vol.01/nr01/a02 (accessed 24 January 2019).

Torre, S. (2018) *Contro la frammentazione. Movimenti sociali e spazio della politica*. Verona: ombre corte.

Tosi, S. and Vitale, T. (2009) Explaining how political culture changes: Catholic activism and the secular left in Italian peace movements. *Social Movement Studies*, 8(2), 131–147.

Tronto, J.C. (1993) *Moral Boundaries: A Political Argument for an Ethic of Care*. New York and London: Routledge.

Tronto, J.C. (2012) *Le risque ou le care?* Paris: PUF.

Tsing, A.L. (2012a) Empire's salvage heart: why diversity matters in the global political economy. *Focaal – Journal of Global and Historical Anthropology*, 64, 36–50.

Tsing, A.L. (2012b) On nonscalability: the living world is not amenable to precision-nested scales. *Common Knowledge*, 18(3), 505–524.

Tsing, A.L. (2015) *The Mushroom at the End of the World: On the Possibility of Life in Capitalist Ruins*. Princeton and Oxford: Princeton University Press.

Zask, J. (2010) Self-gouvernement et pragmatisme; Jefferson, Thoreau, Tocqueville, Dewey. *Etica & Politica/Ethics & Politics*, 12(1), 113–133.

7 Urban green communities

Towards a pragmatic sociology of civic commonality in sustainable city-making

Anette Gravgaard Christensen, Jakob Laage-Thomsen, and Anders Blok

Introduction: urban green communities in question

In recent years, many new civic environmental initiatives have taken root in cities across the world, including in Denmark, in the shape of neighbourhood-based gardens, local food collectives, urban beekeeping groups, tree-planting campaigns, biodiversity efforts in urban green-spaces, and other kindred 'green' activities. On social media platforms and in press coverage and municipal reports, such civic initiatives are often (self-)described in the language of (green) 'community' (*fællesskab* in Danish). As such, they evoke the ideals and the ideology of a certain quality of sociality that is highly valued in Danish (as indeed in other) societies. Moreover, in ways that seem less than coincidental, such descriptions also hark back to the etymological roots of *fællesskab* as oriented to the practice of sharing sheep or cattle together in a customary lawful body that is anchored in a shared place (a *fælled* or 'common') (Bruun 2011). For these reasons, we should ask: how exactly do today's urban green communities shape and constitute themselves?

To speak of an urban green community, however, is to evoke an essentially contested term from the long history of socio-political thinking writ large, that is, the very notion of community. Few would conceivably contest Steven Brint's (2001, 2) assertion that the community concept (in Tönnies' lineage from the *gemainschaft-gesellschaft* distinction) 'has largely failed to yield valuable scientific generalizations', including in the field of urban community studies. On the other hand, we may still agree with Vered Amit (2010) that community is 'good to think with'. While this term is inherently vague, it conjures up core questions as to how social relations form and how social coordination happens under different circumstances. In other words, the notion of community usefully alerts us to questions such as: how, when, where, and why do people come together, what are the terms of their engagement, to what extent are they able to establish and perpetuate a coordinated effort, and how do they feel about it? Such questions clearly matter for civic-green initiatives seeking sustainability in contemporary cities.

In this chapter, we engage with existing studies of city-based environmental politics in order to clarify the analytical premises whereby urban green communities mark out a relatively distinct type of present-day civic green engagement. We believe that such an exploration is motivated by those large-scale transformations underway in cities, as aspirations of sustainability, low-carbon transition, climate resilience, and other kindred environmental goals have long since entered the realm of urban planning and policy institutions as core legitimation narratives (Wachsmuth and Angelo 2018). Meanwhile, relative to the broad-scale mobilisation of urban ecological movements and experiments in the 1970s and 1980s, widespread analyses cast today's aspirations for sustainable city-making as technocratic and lacking in civic participation (e.g. Jamison 2008; Béal 2012). To the extent that urban green communities manifest an emerging form of civic-democratic engagement with the sustainable city, such diagnoses may need revision.

In order to clarify the issues, we review an increasingly inter-disciplinary literature on urban green communities, paying attention to how the language of community is mobilised to rather divergent theoretical, normative, and empirical effects. This ranges from attempts to embrace community's (purported) sense of place as integral to environmental betterment (e.g. Reid and Taylor 2003) to dismissals of the term's ideological use as empty, 'phatic' communication in power-laden discourses of government and industry (e.g. Aiken 2017). More specifically, we identify three main strands of existing conceptualisations: first, an ecological reassertion of long-standing interests in community as a matter of attachments to one's neighbourhood; second, a critical literature stressing how discourses of community functions as a strategy of neo-liberal urban environmental governance; and third, a more political-theoretical notion of imagined risk communities of civic 'green' solidarity and responsibility.

This conceptual mapping serves as a springboard for our core argument: that in order to adequately grasp the particularity of present-day urban green communities, we need to think across and co-articulate these otherwise separated-out senses of community. We achieve this, in the latter part of the chapter, by showing how French pragmatic sociologist Laurent Thévenot's notion of 'commonality in the plural' and its distinction of familiar, planned, and justifiable engagements in particular allows us to integrate the three existing strands of studies into a more comprehensive conceptual architecture. What this architecture allows in relation to existing frameworks is the placement of analytical and empirical emphasis not on three senses of community separately, but rather on the ways in which *any particular* civic urban greening group must co-articulate, or 'co-compose', these forms of commonality *together* to achieve their coherence and collective efficacy over time.

This argument, in effect, constitutes a proposed research agenda which would allow, in subsequent work, a more systematic and comparative study of present-day civic engagements with urban sustainability politics than what is currently on display. For present purposes, we draw on a case study conducted via digital methods (Rogers 2013) that explores commonality-making in a single urban green community, dubbed here Transition Bromst (a pseudonym), which operates an edible forest garden just outside the second largest city in Denmark and is affiliated with the wider Transitions Town network. Evoking this case in relation to our reading of the three strands of literatures allows us to gradually build up a more comprehensive picture of the group's specific composition of sustainability politics than what would be possible from within any one of these strands taken separately. As such, rather than a full empirical account, the case serves here an illustrative purpose intended to indicate a more encompassing agenda for future studies.

Community I: reasserting place attachments

One obvious starting point for this inquiry is with the proliferating sites of urban gardening commonly referred to as 'community gardens' (or *fælleshaver*, in Danish). In many ways, urban community gardens embody a set of key traits that we take also as characteristic of the wider notion of urban green communities. First, they are tied to a specific geographical location in the city, that is, a place whose composition and atmosphere they help alter or shape to varying degrees. Thereby, second, they typically express an active engagement on the part of neighbourhood or other urban inhabitants with shaping their socio-material surroundings in 'greener' directions. As such, third, they embody spaces in which ecological issues, broadly construed, become amenable to forms of everyday, communal, and practice-based civic-political engagement. This is a form of engagement, in turn, that is hard to disentangle from aspects of sociation as such, given that people may seek it out precisely for the sense of community it invokes and portends.

Unlike (purely) imagined or virtual communities (see Brint 2001), then, urban civic-green groups such as gardens, food collectives, and biodiversity efforts stand out in part because they tend to combine aspects of emplaced, value-elective, and activity- or practice-based communities. The precise ways in which they do this, however, can be very different. According to existing surveys, both cross-nationally and within specific cities, publicly accessible gardens are places that are valued for different reasons by different gardeners. While community gardens are sources of improved diet and nutrition in some contexts, such as in poorer Black neighbourhoods in the United States (Kurtz 2001), recreation and relaxation rank higher for people in other situations, while neighbourhood reclamation, social relations, or enhanced proximity to nature do so in yet other settings. Indeed, community gardens may well be popular places in part due to the diversity of motivations and values they accommodate.

More fundamentally, as Hilda Kurtz (2001, 668) notes from a study in the city of Minneapolis, Minnesota, in the United States, community gardens 'serve as tangible and dynamic arenas in which urban residents construct and reinterpret over time the character and meaning of both urban garden, and urban community'. To show this, Kurtz studies three quite different community gardens in the city. One is an older allotment-style garden with individual plots, where people come to grow vegetables in solitude. By contrast, the two other gardens seek to draw upon and build up community, albeit in different ways. One garden is fenced off and reserved for neighbourhood residents' supplemental food provision, with the implication that the boundaries of community are actively renegotiated in sometimes contested ways (along place-based and ethnic lines). The last, a publicly accessible garden, is unenclosed and tended communally by volunteers, which means that it brings together a wider and more loosely defined group from beyond the immediate neighbourhood.

Kurtz's study is useful in part because it alerts us to the historical layers of civic engagement with urban green spaces, as embodied in allotment gardens from the early industrialisation era, as well as to tensions that may arise between more 'place-based' and 'interest-based' gardens (Firth *et al.* 2011). Specifically, her third example, a publicly accessible garden, valuably shows how such categories blur in many present-day community gardens, since these bring together users from within and outside the immediate neighbourhood with attendant variations in motivations and values. Following the study conducted by Pim Bendt *et al.* (2013) from Berlin, we may define this latter style of garden – the one we associate here with the notion of an urban green community – as 'public-access' community gardens. These are 'gardens that are open for anyone at all times, collectively managed by various interest groups in civil society, and in which formal obstacles for immediate participation by the public are absent to low' (Bendt *et al.* 2013, 19).

To analyse the activities conducted in these public-access gardens, Bendt and his colleagues propose that we think in terms of social learning and draw on the concept of 'communities of practice' first introduced (for other purposes) by Jean Lave and Etienne Wenger. Communities of practice are defined by the simultaneous presence of a joint enterprise (e.g. managing a garden), forms of mutual engagement through which people bond (e.g. the actual gardening), and a shared repertoire of rules, jargon, and metaphors that enable a community to reflect upon itself and pursue shared goals. In a community garden context, they suggest, this shared repertoire includes a range of material and ideational elements, from shared tools and composting systems to narratives of solidarity and communal lifestyles (Bendt *et al.* 2013, 23). On this point, however, the notion of communities of practice resonates with, but also differs from, the conceptualisation of urban green communities that we are working towards in the pragmatic-sociological vein.

Most importantly, we argue that the notion of communities of practice renders ambiguous how we should analyse the relationship of community gardens and other urban green communities to issues of neighbourhood and place-based sociality (which are otherwise central to Kurtz's and others' studies). To Bendt and his colleagues (2013), this relationship pertains to two aspects of social learning: 'the politics of space' and 'local ecological conditions'. Regarding politics of space, they highlight how urban gardening groups invariably end up in negotiations and sometimes struggles with the local government, often over issues of commercial developments projected by urban planning. Regarding local ecological conditions, in contrast, they highlight how embodied experiences with gardening lead to greater knowledge of local soil quality, shade and wind patterns, and climatic heat levels. Here, by implication, place is set in opposition to the wider urban space and reduced to its ecological, rather than its social, qualities.

In other words, while claiming that public-access community gardens foster a 'sense-of-place' (Bendt *et al.* 2013, 28), it is somewhat unclear what notion of place is at work in Bendt *et al.*'s invocation of communities of practice. This is an important question since, as Trentelman (2009) points out, there is no simple way in which notions of community map onto or cohere with ideas of 'place attachment', which are understood broadly as those affective relations of belonging, identity, or dependence that may emerge between people and their immediate natural or built environments (see also Jørgensen *et al.* 2016). Conversely, a personal sense of place-based attachment and belonging, for instance, when one feels 'at home' in a neighbourhood, is not necessarily concomitant with any palpable sense of joint commitment or collectivity (Amit 2010, 361). On the other hand, Kurtz's study implies that exactly such issues are likely to emerge in any specific joint enterprise around a public-access community garden since divergent attachments to and orientations towards the immediate urban neighbourhood shape the interactions at work.

Questions of place attachment are also prominent in the Transition Bromst group, which like other Transition Towns has worked hard to attract local members and develop practical initiatives to 'transition' their town in a more sustainable and just direction. Indeed, as Gerald Aiken (2017) has shown, Transition Town activities such as these are steeped in the language of community (*fællesskab*) that are meant to articulate a desire for local-based collective action. In Bromst, as noted, the primary local initiative has been to establish an edible forest garden based on permaculture models. This garden requires sustained care and maintenance work; as such, it becomes a shared reference point for existing group members and attracts new local residents. Aside from tending the garden together, people meet to grill a meal and eat communally, host parties, and share the garden's harvest (Figure 7.1). Over time, members increasingly share photographs of such 'intimate' scenes of their socialising in their Facebook group. The garden, in short, increasingly serves as a valued meeting-ground or 'common-place' (Thévenot 2014) for both members and local passers-by to cultivate personal affinities to each other, to the plants, and to the place as such.

To summarise: existing studies of urban community gardens, as well as our own case study, pose the challenge of combining aspects of emplaced, value-elective ('green'), and practice-based community in ways that are still not captured well in the literature. We argue that this challenge is based in part on inherent ambiguities of the very notion of 'community', as this term is deployed in overlapping yet variable senses across different studies. Rather than solve this issue via a purely typological approach (à la Brint 2001), we argue that such ambiguities may be clarified once we explore how *other*, more critical and political-theoretical frameworks cast 'urban green communities' in substantially

Figure 7.1 Communal dinner around the fireplace in the Transition Bromst edible
  garden, 2017.

Source: reprinted with permission; photographer anonymous.

different lights, bringing out other aspects of the phenomenon. Having taken
such critical-theoretical and political-theoretical points on board, we return
later to a discussion of ways to integrate these aspects of commonality-making
via pragmatic sociology.

Community II: neo-liberal governance strategies

At least since the 1980s, the notion of community has formed a central axis of
more-or-less globalised urban policy formations and practices having to do
with neighbourhood and wider urban change. Here, a range of urban problems
– from planning to crime and from social mobility to ethnic differentiation –
has come to be expressed among diverse urban policy actors in the register of
neighbourliness or community life (Barnett and Bridge 2017, 1199). Right
down to seemingly trivial concerns with such issues as dog fouling and car
parking, notions of community and neighbourhood serve in such contexts as
expressions of the lived experiences of negotiating the boundaries between
autonomy and collective rule-setting under conditions of urban density. In
short, 'community' has long since attained central importance in diverse

practices of urban governance in ways that impinge strongly also on urban green communities.

In the realm of community gardens, Marit Rosol (2010, 2012) shows how discourses of community volunteering in Berlin are connected to neo-liberal local governance strategies. According to Rosol, the aim of connecting volunteering to such strategies is to activate and involve citizens and other civil society actors in urban governance in order to outsource former local state responsibilities by using volunteers, for example, to initiate and maintain urban green-spaces. Rosol (2010, 555) locates a set of existing models for creating community gardens that are spread across an axis of civic-political cooperation and conflict. These models range from self-organised citizen projects actively supported by urban planners, via grass-roots projects developed in open defiance of planning-based regulations, and gardens located in urban renewal areas that are initiated mainly by urban administrators themselves. It is thus far from obvious, at the outset, where the boundaries are to be drawn between civic engagement and volunteering on the one hand, and institutional urban governance on the other, given the partly shared and partly divergent agendas.

Against this backdrop, Rosol seeks to supplement the dominant 'emancipatory' interpretation of community gardens, which portrays them 'as both critiques of, and actually existing alternatives to, traditional state-provided open green spaces' (Rosol 2012, 240). In such narratives, community gardens are sites in which citizens struggle to assert their right to use and shape the city. In Rosol's view, however, this narrative does not fully grapple with the reality of local state budgets under pressure, whereby the invocation of 'community' becomes that of the activating state seeking to embrace citizens as potential resources in community-based volunteer groups. From a local administrator's perspective, the collective effort of cleaning and greening an empty lot to establish a community garden produces positive effects for 'the community', since the effort is seen as 'an important means of creating community identity and thereby responsibility for and "stabilization" of the area'. In addition, Rosol (2012, 245) suggests such strategies could make the urban area in question more attractive to tax-paying, middle-class families.

Importantly, however, Rosol's study finds that the local state in Berlin does *not* succeed in establishing a shared conviction with civic gardening groups regarding the necessity of taking over maintenance responsibility for 'the green sector' as such. Instead, she finds that participants' motives for volunteering specifically in community gardens, rather than in city parks, are connected to a variety of what she calls 'personal reasons'. In contrast to volunteering in the public parks, community gardens are self-organised and 'community managed, that is, they are collectively designed, built, and maintained by local residents' (Rosol 2012, 243). Such self-determination, Rosol argues, is an important factor in the more-or-less long-lasting commitment of volunteer residents, who do not want to be 'just cleaning, picking up garbage' in the city parks just to relieve what are considered local state responsibilities (Rosol 2012, 247). In other words, despite intense discourses of 'community', the local state in this instance does not succeed in fully defining the responsibilities of these urban green groups and its full outsourcing strategy fails.

We argue that Rosol's study is important because it helps shine analytical light on core differences in motivations for supporting community gardens and other urban green communities and on the various power relations in play. However, her largely neo-Marxist framework does not fully explain the intricate dynamics whereby the local state fails to harness the resources of civil society in green-space maintenance. Nor does it quite capture how such strategic state attempts serve inadvertently to activate a network of self-organising community gardeners, which in turn do not fully succeed, either, in making their long-term 'green' wishes count in urban planning. Simply put, Rosol's study concludes that the economic interest of the local state does not align with the diversity of interests of the community gardeners. Yet, rather than being a satisfactory explanation, this statement arguably calls for further clarification itself, since actors 'interests' should be seen as shaped in power-laden yet interactionally open dynamics of democratic contention.

Working broadly in this direction, Ane Grubb and Lars Skov Henriksen (2018) analyse the changing civic landscapes in Denmark and similarly identify a 'welfare discourse' whereby state actors strategically seek to enhance voluntary contributions in solving welfare challenges. Unlike Rosol, however, this situation leads them to problematise an assumption that is often found in civil society discourse: the assumption that volunteering, no matter the form, inherently contributes to the creation of civic virtues and that any voluntary organisation is a potential 'school of democracy' (Grubb and Henriksen 2018, 5). In order to refine their inquiry beyond this assumption, Grubb and Henriksen turn to Paul Lichterman's and Nina Eliasoph's (2014, 809) concept of *civic action*, defined as where 'participants are coordinating action to improve some aspect of common life in society, as they imagine society'. Invoking this concept serves here to better delineate the democratic potential of civic participation and to specify what makes action 'civic' in order to be able to establish whether interaction qualifies as civic action and which organisational forms enable or constrain it.

Grubb and Henriksen specifically analyse a new type of volunteering, which they label 'program volunteering', which differs from the classic membership-based voluntary organisation. This concept captures types of volunteer initiatives, financed by a publicly or privately sponsored programme, which aim to produce services that target third-party beneficiaries. Using a case study of this type of programme volunteer organisation (in the educational sector), Grubb and Henriksen find limited space for civic action. The organisational setting in question combines hierarchical and business managerial styles with tasks and practices of the volunteers that are standardised and pre-defined, which leaves only marginal room to 'discuss and interpret the wider social and political context of the problem they were trying to solve' (Grubb and Henriksen 2018, 8). As a result, volunteers for the organisation under study did not see their role as a shared practice to improve life for the target-group they were helping, which meant that their volunteer efforts fail to establish what their organisation's CEO otherwise terms 'community'. The volunteers, in short, were constrained and not able to transform their engagement into civic action.

This type of municipal programme volunteering approach can also be found in the green sector, and indeed shows up in our Transition Bromst case study, albeit in ways that inter-articulate with more classic forms of volunteering and grass-roots activism. Here, the approach is framed around attempts of this initiative-cum-voluntary association to compete with other stakeholders for space and resources in the local political economy. Notably, it took two years of petitioning and negotiating with the municipality for Transition Bromst to gain access to a municipal plot of land on which to establish the edible forest garden (Figure 7.2). During this time, the group actively sought to mobilise the local citizenry to influence and put pressure on municipal environmental plans, thus acting as a concerned grass-roots public. Later on, however, as the garden proved viable, the municipality sought to re-embed the garden into a wider 'green partnership', which was set up with the Transition Bromst group and a national NGO in order to provide school gardens for local kids. While this municipal action expressed the voluntary groups' 'success' to some extent, such re-framing also potentially weakens its civic space of action.

Figure 7.2 General board meeting in Transition Bromst, discussing, for example, plans for negotiating with the municipality, 2017.

Source: reprinted with permission; photographer anonymous.

To summarise: based on our own and existing research on the ambiguous power-spaces of urban governance, we argue that a notion of civic action as 'coordinated interaction around a mission to improving common life' (Lichterman and Eliasoph 2014, 810) needs to be at the core of analysing urban green communities and the more-or-less civic spaces they entail. Here, in navigating the increasingly pluralistic scene of sustainable city-making, the concept of civic action helps demarcate urban green communities from other types of initiatives. In doing so, we also acknowledge that urban green initiatives constitute arenas of constant civic-political negotiation, both internally and vis-à-vis external state and other institutions, and that analytical sensitivity to the unfolding of such negotiations is key to the civic-political character of the community they embody. With this in mind, we turn now to 'imagined (green) communities' as the third and last strand of existing literature holding direct relevance to urban green communities.

Community III: imagined urban-green solidarities

In the tradition of urban studies, cities have long been cast as not only sites of place-based everyday community life, nor simply as a particular scale, vector, or diagram of government. Rather, cities and urban coexistence also pertain to more fundamental questions of ethical-political solidarities and basic parameters of political community. Borrowing from Benedict Anderson's (1983) famous rendering of the nation-state as an 'imagined community' of mutual affinity and cooperation among people who most likely never meet each other, scholars have debated what might be different about, and perhaps specific to, imagined urban communities (e.g. Magnusson 2014). While it is in many ways more unwieldy and open-ended than the two other literatures, we suggest that this more political-theoretical approach to the question of urban green communities proves important to our effort to capture aspects of these communities that are otherwise difficult to articulate.

Ash Amin's (2006) ruminations on the conditions of the 'good city' and how to study it (Amin 2007) serve as a valuable first approximation. A city, he suggests (Amin 2007, 109), is a political 'community of communities' in the sense of being 'a site of manifold mobilizations' which involve 'solidarities and antagonisms that extend well beyond the city'. This use of the term 'community' is thus no longer associated with spatial or place-based continuity but rather casts the city as a relational setting that potentially brings the distant near in affectively important ways. Moreover, Amin (2007, 107f.) argues, a set of non-place-specific issues, commitments, and concerns increasingly extend beyond 'the human' into issues of how urban-based technologies and natures become sites of socio-political struggle as well as new forms of affect and solidarity. Analyses of urban community, in short, must now come to see questions of nature and the environment as belonging to the inside, rather than the outside, of the civic-political life of cities.

While Amin's work is inspiring, his suggestions as to *how* nature(s) and the environment matter to the formation of urban imagined communities are somewhat short on specifics. More adequate suggestions on this question can be found in the work of Ulrich Beck and his colleagues on so-called 'cosmopolitan communities of climate risk' (Beck *et al.* 2013), including how this latter notion pertains to contemporary climate-political ambitions in world cities around the globe (see Blok 2016). Drawing explicitly on Anderson's work, Beck *et al.* (2013 2) define these risk communities as 'new transnational constellations' that arise 'from common experiences of mediated climatic threats' and shared reasoning on responsibilities, and thereby potentially enable 'collective action' and 'international norm generation'. One example, they argue, is the emergence of interurban networks such as C40, where urban planning and policy-making elites establish, coordinate, and circulate new norms of low(er)-carbon urban development.

While attuned to the potential for new conflicts over resources, this notion suggests that *one* type of urban response to the challenges of climate change takes the shape of new imagined 'green' communities of mutual policy coordination. In line with Amin, such risk communities involve solidarities and antagonisms that extend beyond the individual city, as mediated largely via commitments to climate change as a globally shared ecological concern and a sphere of urban collective action. In turn, forums such as C40 serve to articulate a set of power-laden narratives when it comes to integrating economic competitiveness and environmental performance into urban sustainability. While civic groups and mobilisations may play into these strategic spaces and may sometimes succeed in pushing for stricter local climate-political goals, the domain of climate risk politics in the city mostly takes its shape around rather technocratic policy commitments to socio-technical interventions in infrastructures of energy, housing, traffic, and waste (Blok 2016).

As such, the extent to which climate risk communities as defined by Beck overlap with or impinge on urban green communities, as we deploy this term here, is quite a contingent question. In so far as urban sustainability politics spans a wide range of arenas and issues – including green-space provision – convergences and divergences are likely to coexist at the level of civic engagement. Hence, for instance, in today's Danish urban context, official government-driven interventions in the name of climate adaptation often also entail enhancing urban green-space via small parks or plant-filled squares, which also comes with ambitions to leverage the resources of volunteers in co-shaping and maintaining such new green-spaces. Here, the articulation of globalised climate risk communities at the localised scale of the urban neighbourhood raises questions about the scope for authentic civic green action that such interventions in fact create, which is in line with the more general dilemmas set out in the previous section. Importantly, however, what exactly is entailed by 'greening' in such instances also remains subject to contention between more technical and more verdant senses of urban nature (Wachsmuth and Angelo 2018).

Turning these questions around, one should ask about the extent to which concerns with otherwise 'distant' climatic changes are *also* part of motivating more localised, practice-based urban green communities. Arguably, the most poignant example of an urban-inflected grassroots environmentalist movement to have taken on board the challenges of climate change and peak oil is the Transitions Town Network to which our case study group belongs. By collectively engaging in edible forest gardening, permaculture courses, and other practical activities, this group also seeks to articulate new forms of local political agency that are tied to the wider and alternative imagined urban community ethos made available by the transnational network. This ethos, in turn, becomes the subject of a variety of so-called 'vision meetings' (Figure 7.3), in which group members actively articulate and justify their shared green (low-carbon, biodiversity) and civic (local democracy, stronger community) moral-political ideals for societal change (Boltanski and Thévenot 2006 [1991]). In short, these meetings are sites in which visions for global environmental and social justice come to sit side-by-side with more vernacular local activities of embodied care for the garden.

Hence, what emerges more generally between dominant urban sustainability policies on the one hand and Transition Town alternatives on the other is a sense of the civic-political ideational and indeed the ideological landscape of city re-imaginations in which contemporary urban green communities operate. This space should itself be seen as highly variegated, in part because it spans from the most localised concerns with *this* valued park, garden, or tree at the one end, to the most globalised concerns with climate change, planetary survival, and North–South inequalities at the other. Where specific urban green

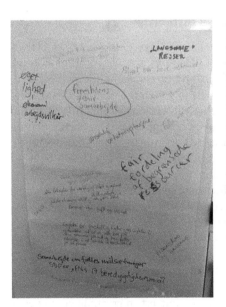
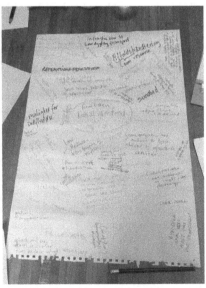

Figure 7.3 Shared output from vision meeting in Transition Bromst, discussing, e.g., 'fair sharing of limited resources' in a global perspective, 2017.

Source: reprinted with permission; photographer anonymous.

communities, say, a certain bird-watch group or a food cooperative, choose to place themselves in this ideational landscape and how they seek to combine their specific concerns into more comprehensive greening narratives is entirely open-ended in empirical terms. For the group in Bromst, for instance, aligning everyday forms of gardening care with wider permaculture principles, not to mention concerns with climatic justice in a global scale, is far from an easy task, subject as it is to ongoing adjustment and negotiation.

To summarise: just as we need to question the meaning of 'urban' and 'community', we simultaneously need to question the *kinds* of imagined urban communities suggested by the notion of civic urban 'greening' politics. In doing so, we should ask new questions about the kinds of 'ecological citizenship' (Wolch 2007) such communities help cultivate in different settings, spanning the range from anti-capitalist subjectivities committed to 'urban commoning' (Bresnihan and Byrne 2015) all the way to 'post-political' imaginations of the environmentally conscious urban consumer fit for a world of green capitalism. In order to articulate such differences, however, we need a stronger analytical vocabulary for analysing moral-political coordination than the notion of 'imagined community' affords by itself. We turn now to discuss the extent to which Laurent Thévenot's (2014) pragmatic-sociological notion of commonality in the plural may prove more analytically fruitful in this respect, allowing us to bring key insights from the three preceding sections – and literatures – on 'community' into a comprehensive framework.

Civic commonality in the plural in the sustainable city

In recent years, extending his framework of a sociology of engagements, French pragmatic sociologist Laurent Thévenot (2014) has articulated an alternative vision of political community under the rubric of commonality in the plural. The following discussion explores the extent to which this notion productively extends *and* recasts the conceptual and empirical tensions located thus far at the level of specifying an analytics of urban green communities. Our basis for doing so is straightforward: Thévenot's conceptual architecture of commonalities, which distinguishes personal affinities to common-places, choice in a liberal public, and plural orders of worth, respectively (see Table 7.1), seems to match in interesting ways the three bodies of literature so far reviewed. In other words, with a slight change in conceptual register from 'community' to 'commonality' (in the plural), we seek to integrate the various analytical tensions into a single, comprehensive model of the dimensions of variation at work in any single urban green community.

Table 7.1 should be read to suggest that, contrary to the discussion so far, which treats the three strands of literature on community separately, Thévenot's framework allows us to analyse how *any* given civic urban greening group needs to coordinate *all* three grammars of commonality together to create and maintain their coherence and collective efficacy. This conceptualisation marks out a space of variation and empirical possibilities that allows us to specify the

Table 7.1 Three grammars of commonality in the plural that are active in an urban green community

Grammar of commonality	Plural orders of worth	Choice among options in a liberal public	Personal affinities to common-places
Communication of agreement and difference	Qualifying concerns into common worth (market, industrial, civic, domestic, opinion, inspiration, green)	Opining and negotiating over choices made available as options open to the public	Diversely associating personal affinities to shared common-places
Assumed engagement	Justification and critique for the common good	Individual and collective plans	Familiarity
Type of 'community'	Imagined political communities (III)	Governance arrangements (II)	Place-based attachments (I)
Transition Bromst case	Combining and compromising green and civic worth	Navigating the civic space of municipal plans and partnerships	Shared care and maintenance of edible forest garden

Source: adapted from Thévenot (2014, 18).

particular ways in which an urban green community co-articulates its common-alities, as well as to start comparing such group-based and wider organisational compositions across cases locally, nationally, and transnationally. We argue that this conceptualisation improves upon existing analytical tools because it allows us to see how present-day urban green communities – such as the Transition Bromst group, our illustrative case – stand out as a relatively distinct socio-historical form of civic-political engagement vis-à-vis the contours of urban sustainability politics.

Here, we can only indicate some key parameters for this wider research agenda. To commence in reverse order, the commonality of plural orders of worth refers here to the idea that civic groups or collectivities, or what we called imagined political communities, establish themselves and undergo change via shifting compromises among a plurality of ideas of the shared common good. In *On Justification* (Boltanski and Thévenot 2006 [1991]), Thévenot, together with Luc Boltanski, identifies six 'orders of worth', each constituting a grammar of legitimate socio-political bonds, while each is also mutually incompatible in their value orientations. The market worth of economic exchange and competi-tion, for instance, is at odds with the civic worth of solidarity and legal equality. Yet, the two can be brought together in a more-or-less durable compromise, as happens routinely in cities under the planning rubric of sustainability (Blok and Meilvang 2015). Amin's urban 'community of communities' can thus be recast from this angle as overlapping spheres of shared or divergent justifications and critiques of the principles of urban ordering.

During the 1990s, Thévenot and his colleagues identified an additional, seventh order of worth associated with environmentalist or green principles and values. In a series of case studies into nature conservation controversies in France and the US, Thévenot *et al.* (2000) distilled the contours of such an emerging green order, which is oriented towards the common good of environ-mental protection, biodiversity, and overall 'naturalness'. Further discussions have sought to consolidate this finding by pointing to the importance of dis-tinctly green forms of justification and critique in contemporary sustainability politics, while also raising questions as to the ambiguities that still surround shared senses of green worth. Hence, for instance, while climate politics rou-tinely translates green worth into calculable forms of carbon budgeting, urban environmental activism often relies instead on staging green-spaces as a source of 'authentic' living (Blok 2013) or as part of 're-localising' economies (Chia-pello 2013). Such 'green' ambiguities correspond to, and allow for a firmer grasp on, the landscape of civic-political re-imaginations in which contemporary urban green communities such as Transition Bromst operate in their diverse ways.

In other words, the commonality of plural orders of worth in the city consists of urban civic groups coordinating their activities around a limited plurality of value orientations considered to embody the common good of all or, conversely, critiquing planning arrangements for failing to do so. Urban green communities, we argue, should thus be analysed in terms of their shared activities of public green justification and critique, including how they seek to forge compromises with other forms of worth. As in the Transition Bromst case, these are likely to include variable senses of civic worth that correspond to different scales of collective commitment: in other words, how the 'imagined green community' is seen to extend outwards from the immediate neighbourhood to the city at large and potentially into transnational spaces of solidarity. Such civic-green compromises, we suggest, constitute one major axis along which urban green communities differ, both within specific cities and across trans-national contexts.

The major advantage of Thévenot's notion of commonality in the plural, however, only becomes apparent once this idea, which pertains to *one* type of commonality-making, is seen and analysed in tandem with the other two types. Hence, in the type of commonality dubbed 'choice in a liberal public', actors orient themselves not according to abstract moral-political values but rather according to a sense of collective self-interest as this comes to be negotiated amongst stakeholders in the context of planned arrangements. This, one might say, is the standard modality of contemporary urban planning, whereby a variety of urban agents – from professionals to local businesses, voluntary associations, and individual users – come to be enrolled in and negotiate over specific plans for urban development (Meilvang *et al.* 2018). As such, this type of commonality, we argue, shapes what we previously called local governance in relation to civic volunteering and thus also the space of (failed) 'alignments' that this relation entails.

In commenting generally on the kind of neo-Marxist approach found in Rosol's study of green volunteering, Thévenot (2012) points to the approach's difficulty in grasping the characteristics of what it calls 'neoliberal' urban governance. Far from simply a matter of market dominance, he suggests, contemporary forms of 'governance by objectives' are marked instead by a strict reliance on ways of standardising and certifying the choices made available for actors in the liberal public. One important illustration of such tendencies, arguably, is 'programme volunteering': here, the goals to be achieved come to be formalised via contracts between volunteers, public–private institutions, and funding agencies, which paves the way for new forms of accountability regimes to which volunteers are subjected (Grubb and Henriksen 2018). In the sustainability domain, tendencies for local governments to seek partnerships with volunteer groups can be read in similar terms, as pointed out in the Transition Bromst case. Here, one must carefully adjudicate what space for manoeuvre for civic action, including critique, emerges in such settings.

The resultant tensions for urban green communities between engaging in public green critique or entering into formalised partnerships with local governments exemplifies what Thévenot (2014) calls the 'sacrifice' involved in pursuing certain engagements over others. At the same time, his model entails that civic groups over time attain characteristic 'group styles' (Eliasoph and Lichterman 2003) by juxtaposing and co-articulating (or 'composing') such plural engagements in specific ways. Hence, much as in the Transition Bromst case, one can imagine an urban green community entering a formal partnership with the municipality over specified policy goals, such as teaching school children about edible plants and vegetables, while it simultaneously struggles to turn its own garden into a platform for wider public critiques of otherwise unsustainable green policies in the city. While such an arrangement may work smoothly in some settings (as it seems to do in Bromst), tensions may come to the fore in others, as civic participants negotiate what degree of 'critical-ness' towards the municipality is compatible with their partnership. In other words, any specific urban green community will continuously be re-negotiating such tensions at the level of their group style.

For the analysis of urban green communities, moreover, Thévenot's third type of commonality, that of personal affinities to common-places, is likely to be a crucial part of any such composite group style. By 'common-place', Thévenot (2014) invokes the classical idea of an important and affectively shared reference point, whether in the shape of a poem, a family house, or indeed a certain green-space in the city that one frequently visits and where one 'feels at ease'. Such highly personal and embodied experiences may in turn form the basis for a specific type of commonality, Thévenot argues, less by way of formal argument (over worth) or the negotiation of interests (in a liberal public) but more at the level of a shared attunement of affects and affinities to specific objects of care and concern (see Blok 2015). As such, this notion resonates strongly with the types of place attachments and their embedding in practice-based civic communities that we discussed as the first strand of literature on urban green communities. Indeed, we suggest that Thévenot's notion helps render such ideas more analytically coherent by specifying the particular type of affinity- and care-based sociality likely to characterise such engagements.

In many ways, then, Thévenot's notion of affinity to common-places helps make sense of what is arguably a key present-day tendency in urban–civic engagement set up around novel civic articulations of place-based ecological concerns in the city. These articulations include new neighbourhood-based urban river advocacy movements (Wolch 2007), local civic initiatives for the recreational restoration of waterways (Karvonen and Yocom 2011), and new forms of 'emplaced resistance' (Centemeri 2017) on the part of green-spaces vis-à-vis large-scale infrastructure developments. In 2017, Copenhagen in Denmark similarly witnessed the emergence of a large-scale civic mobilisation for the preservation of an urban green-space known as *Amager Fælled* ('Amager common'), which was driven in part by concerns for the embodied, practice-based forms of appreciating nature and biodiversity that this place affords. In all of these cases, we might say, with Thévenot, that personal affinities to certain green-spaces positioned as common-places are likely to have formed an important backdrop to wider, civic-political mobilisations.

This does not mean, however, that the forms of shared affinities at work in various urban green communities, including the edible forest garden in Transition Bromst, necessarily attain such a publicly visible and politicised expression. On the contrary, it seems rather characteristic of many such initiatives, including those across Denmark, that their overall style of commonality tends to prioritise exactly the more informal and personally invested familiar engagements with a place and its ecological qualities that Thévenot articulates in the notion of the common-place. In other words, only under specific conditions do urban green communities put such shared affinities into sustained contact with, and *explicitly* compose them vis-à-vis the more formal styles of negotiation characteristic of liberal-public partnerships on the one hand and the more morally demanding styles of green critique characteristic of civic-political debate in the urban public sphere on the other hand. As the Transition Bromst case shows, however, such explicit compositions do happen – and, as such, the way specific urban green communities put the three types of commonality together over time becomes defining of key variations for analysis.

To summarise: in order to lend clarity to the notion of urban green communities, it helps to shift the conceptual register from 'community' to 'commonalities' (in the plural), as this is elaborated in pragmatic sociology. Whereas the literature on community has tended to generate new typologies and perspectives, Laurent Thévenot's framework allows us to treat the core dimensions of community-making as empirical variables along a set of well-defined, internally coherent analytical dimensions of how commonality is forged in specific group settings. This framework, in turn, will afford both in-depth case studies in specific contexts as well as comparative inquiries into the various styles of urban green communities in different local, national, and world-regional settings. As such, beyond the single case analysed in this chapter, the analytical framework will allow analysts to place the emerging phenomenon of urban green communities in the wider context of sustainable city-making. Thereby, in time, it will lend weight to the intuition that what we are witnessing may mark an important shift in civic green engagement and mobilisation – one oriented to its localised, practical, and communal emplacement in the city.

Conclusion: new civic landscapes of urban greening?

In this chapter, we respond analytically to the recent surge of academic and practical interest in urban green communities in order to ask what precise senses and styles of civic-political 'community' (or *fællesskab*, in our Danish setting) they may be said to embody. We do so based on the idea that neighbourhood-based gardens, local food collectives, urban beekeeping groups, and kindred forms of emplaced, practice-based, civic-green engagements with the city and its ecologies suggest themselves as a relatively distinct, novel, and increasingly important form of civic-democratic participation in sustainable city-making. Such initiatives, we suggest, seem sufficiently allied in their styles of civic action to warrant the shared conceptual designation (*as* 'urban green communities'). Yet, as reviewing the blossoming interdisciplinary literature has helped reveal, these initiatives are also likely to exhibit wide variations in forms of sociability and civic-green commitments. A viable analytical approach, we argue, must be able to embrace both facets.

In other words, before being able to fully corroborate empirical claims in this domain, we need a better analytic sense of how urban green communities differ as well as what nevertheless sets them apart from other types of green volunteering or environmental movements. Towards this twin aim, we suggest following recent developments in pragmatic sociology, as epitomised in Laurent Thévenot's (2014) work on engagements and commonalities in the plural. This framework, as we have shown, is capable of integrating insights pertaining to otherwise largely disconnected literatures on (urban) 'community', thereby allowing us to maintain a fine-grained view toward diverse group styles among urban green communities while at the same time gaining more analytical precision on the conceptual dimensions at work. As such, the framework is well suited for both case study and comparative research at various scales of granularity, from within-city variations to broader patterns of cross-national difference.

What makes Thévenot's framework particularly suitable for this purpose is the attention it allows us to pay to the efforts of urban green communities to cultivate at the same time a set of quite informal place- and practice-based attachments and more public forms of justification and critique, even as they negotiate their degrees of partnership with local authorities. Indeed, we argue that this simultaneity of divergent forms of civic engagement and commonality-making – and, not least, the centrality of what Thévenot calls personal attachments to common-places to their shared group styles – may well form an important distinguishing trait between current-day urban green communities and the urban ecological movements of the 1970s and 1980s (Jamison 2008). Compared with the latter, urban green communities so far fail to register quite the same public-critical force. However, while often less visibly and openly antagonistic, their cultivation of shared, embodied, green sensitivities and mutual attunements may well become a strong force in a variety of new 'emplaced resistances' (Centemeri 2017) to urban development as usual, in ways that will shape sustainable city making efforts in the coming years.

As such, urban green communities must ultimately be situated in a changing civic-democratic landscape of urban greening and sustainability – one usually portrayed as suffering in recent years from marked tendencies towards technocratic and professional dominance over planning and policy-making (e.g. Jamison 2008; Béal 2012). We suggest that urban green communities should be seen to constitute one of the more interesting counter-currents in recent years, one that is oriented to fostering a different set of urban commonalities that is more in tune with diverse, place-based ecological attachments in the city (see Blok and Meilvang 2015). While acknowledging this, however, our Thévenot-inspired framework simultaneously allows us to question the terms on which these civic groups must negotiate such informal attachments at the level of their 'rights' to places, vis-à-vis local planning authorities who are more inclined towards standardised forms of governance by objectives. Here, the democratic qualities of sustainable city-making policies should be judged as much by the active, civic contestability of such objectives and the green commonalities they entail as they should by any formalised procedures of public participation (see Meilvang *et al.* 2018).

We end on this note: the proximity that exists between Thévenot's stress on the importance of common-places and the etymological roots of *fællesskab*, the

Danish equivalent of 'community', with which we began our inquiry. *Fællesskab*, as we noted, stems from the pre-modern practice of raising sheep or cattle together in a shared place of grazing, a *fælled* or 'common'. As recent civic-political mobilisations around the so-called *Amager Fælled* in Copenhagen made clear to this city's inhabitants, urban green commons may attain new qualities today as an urban green common-place, a place of diverse-yet-shared affinities to the city and its rich human and non-human ecologies. We suggest that, in a variety of ways, this city's many scattered and small-scale urban green communities formed here an important civic backdrop to a large-scale event of emplaced resistance to what was seen as the un-sustainable planning politics of the local government. While 'community' in the singular may in the end be too much to strive for and not quite what a pluralistic democracy needs, for both academic and democratic reasons we ought to pay more attention to how a growing variety of urban green communities are already building more viable commonalities in relation to our urban commons.

References

Aiken, G.T. (2017) The politics of community: togetherness, transition and post-politics. *Environment and Planning A*, 49(10), 2383–2401.

Amin, A. (2006) The good city. *Urban Studies*, 43(5–6), 1009–1023.

Amin, A. (2007) Re-thinking the urban social. *CITY*, 11(1), 100–114.

Amit, V. (2010) Community as 'good to think with': the productiveness of strategic ambiguities. *Anthropologica*, 52(2), 357–363.

Anderson, B. (1983) *Imagined Communities: Reflections on the Origin and Spread of Nationalism*. London: Verso.

Barnett, C. and Bridge, G. (2017) The situations of urban inquiry: thinking problematically about the city. *International Journal of Urban and Regional Research*, 40(6), 1186–1204.

Béal, V. (2012) Urban governance, sustainability and environmental movements: post-democracy in French and British cities. *European Urban and Regional Studies*, 19(4), 404–419.

Beck, U., Blok, A., Tyfield, D., and Zhang, J.Y. (2013) Cosmopolitan communities of climate risk: conceptual and empirical suggestions for a new research agenda. *Global Networks*, 13(1), 1–21.

Bendt, P., Barthel, S., and Colding, J. (2013) Civic greening and environmental learning in public-access community gardens in Berlin. *Landscape and Urban Planning*, 109(1), 18–30.

Blok, A. (2013) Pragmatic sociology as political ecology: on the many worths of nature(s). *European Journal of Social Theory*, 16(4), 492–510.

Blok, A. (2015) Attachments to the common-place: pragmatic sociology and the aesthetic cosmopolitics of eco-house design in Kyoto, Japan. *European Journal of Cultural and Political Sociology*, 2(2), 122–145.

Blok, A. (2016) Urban climate risk communities: East Asian world cities as cosmopolitan spaces of collective action? *Theory, Culture & Society*, 33(7–8), 271–279.

Blok, A. and Meilvang, M.L. (2015) Picturing urban green attachments: civic activists moving between familiar and public engagements in the city. *Sociology*, 49(1), 19–37.

Boltanski, L. and Thévenot, L. (2006[1991]) *On Justification: Economies of Worth*. Princeton: Princeton University Press.

Bresnihan, P. and Byrne, M. (2015) Escape into the city: everyday practices of commoning and the production of urban space in Dublin. *Antipode*, 47(1), 36–54.

Brint, S. (2001) *Gemeinschaft* revisited: a critique and reconstruction of the community concept. *Sociological Theory*, 19(1), 1–23.

Bruun, M.H. (2011) Egalitarianism and community in Danish housing cooperatives: proper forms of sharing and being together. *Social Analysis*, 55(2), 62–83.

Centemeri, L. (2017) From public participation to place-based resistance. Environmental critique and modes of valuation in the struggles against the expansion of the Malpensa airport. *Historical Social Research/Historische Sozialforschung*, 42(3), 97–122.

Chiapello, E. (2013) Capitalism and its criticisms. In P. du Gay and G. Morgan (eds), *New Spirits of Capitalism?*. Oxford: Oxford University Press, 60–81.

Eliasoph, N. and Lichterman, P. (2003) Culture in interaction. *American Journal of Sociology*, 108(4), 735–794.

Firth, C., Maye, D., and Pearson, D. (2011) Developing 'community' in community gardens. *Local Environment*, 16(6), 555–568.

Grubb, A. and Henriksen, L.S. (2018) On the changing civic landscapes in Denmark and their consequences for civic action. *Voluntas*, 1–12. Available at: https://doi.org/10.1007/s11266-018-00054-8 (accessed 9 January 2019).

Jamison, A. (2008) Greening the city: urban environmentalism from Mumford to Malmö. In M. Hård and T.J. Misa (eds), *Urban Machinery: Inside Modern European Cities*. Cambridge, MA, MIT Press, 281–298.

Jørgensen, A., Knudsen, L.B., Arp, M., and Skov, H. (2016) Zones of belonging: experiences from a Danish study on belonging, local community and mobility. *Geoforum Perspektiv*, 15(29), 1–28.

Karvonen, A. and Yocom, K. (2011) The civics of urban nature: enacting hybrid landscapes. *Environment and Planning A*, 43(6), 1305–1322.

Kurtz, H. (2001) Differentiating multiple meanings of garden and community. *Urban Geography*, 22(7), 656–670.

Lichterman, P. and Eliasoph, N. (2014) Civic action. *American Journal of Sociology*, 120(3), 798–863.

Magnusson, W. (2014) The symbiosis of the urban and the political. *International Journal of Urban and Regional Research*, 38(5), 1561–1575.

Meilvang, M.L., Blok, A., and Carlsen, H.B. (2018) Methods of engagement: on civic participation formats as composition devices in urban planning. *European Journal of Cultural and Political Sociology*, 5(1–2), 12–41.

Reid, H. and Taylor, B. (2003) John Dewey's aesthetic ecology of public intelligence and the grounding of civic environmentalism. *Ethics and the Environment*, 8(1), 74–92.

Rogers, R. (2013) *Digital Methods*. Cambridge, MA: The MIT Press.

Rosol, M. (2010) Public participation in post-fordist urban green space governance: the case of community gardens in Berlin. *International Journal of Urban and Regional Research*, 34(3), 548–563.

Rosol, M. (2012) Community volunteering as neoliberal strategy? Green space production in Berlin. *Antipode*, 44(1), 239–257.

Thévenot, L. (2012) Law, economies and economics. New critical perspectives on normative and evaluative devices in action. *Economic Sociology*, 14(1), 4–10.

Thévenot, L. (2014) Voicing concern and difference: from public spaces to commonplaces. *European Journal of Cultural and Political Sociology*, 1(1), 7–34.

Thévenot, L., Moody, M., and Lafaye, C. (2000) Forms of valuing nature: arguments and modes of justification in French and American environmental disputes. In M. Lamont and L. Thévenot (eds), *Rethinking Comparative Cultural Sociology*. Cambridge: Cambridge University Press, 229–272.

Trentelman, C.K. (2009) Place attachment and community attachment: a primer grounded in the lived experience of a community sociologist. *Society & Natural Resources*, 22(3), 191–210.

Wachsmuth, D. and Angelo, H. (2018) Green and grey: new ideologies of nature in urban sustainability policy. *Annals of the American Association of Geographers*, 108(4), 1038–1056.

Wolch, J. (2007) Green urban worlds. *Annals of the Association of American Geographers*, 97(2), 373–384.

8 The Stop Wasting Food Movement as a community of potentialities

Simon Lex and Henrik Hvenegaard Mikkelsen

Introduction

In the struggle towards a sustainable future, one can identify two general strategies. The first, which can be termed the objective approach, operates on a larger, structural, and systemic level. Here the road to sustainability is paved with technological innovation and large-scale political decision-making. What is commonly emphasised in this approach is, for instance, the political use of subsidisation, taxation and 'nudging', in order to encourage citizens to engage in a 'greener' form of consumption (Thaler and Sunstein 2008). We also find in this approach the ambition to create 'smart cities' where energy production and distribution operate automatically in vast digital systems. In other words, the objective approach does not rely solely on the sense of responsibility and active exercising of restraint of the individual consumer in his or her everyday life. The second strategy, which could be termed the subjective approach,[1] involves the opposite: it places the agency of the individual at the centre of the transition to sustainability by means of the principle that the individual needs to reach a heightened awareness about his or her patterns of consumption. In view of this, any form of sustainability rests upon the crystallisation of knowledge of the impact that people have on the world as individuals embedded in larger communities.

In this chapter we engage with the latter strategy, which lies at the core of many of the activities carried out by various components of environmental movements. Based on our study of the Stop Wasting Food Movement (SWFM) in Denmark, we discuss how individuals with a common pro-environmental cause engage in digitally structured communities. We believe that the case provided by the SWFM reveals an often-overlooked aspect of community, namely the importance of the potential of the notion – or communicated message – that its members are part of a trajectory that may lead to a more desirable future. We suggest that while such potentials may (or may not) be realised in the future, the use of them in the present incites us to imagine the potential as a critical stratum of social reality (Mikkelsen 2017).

Methodologically, the chapter rests on an ethnographic study of the SWFM. Lex has conducted digital ethnography following the development of the movement on social media platforms such as Facebook and LinkedIn for a period of years (2016–2018). Furthermore, through this period, Lex interviewed the founder, Selina Juul, and organisations that support the SWFM. Drawing on this project, we explore how SWFM has succeeded in expanding its initiatives from a preliminary idea to a nationally recognised programme in the last ten years. This was achieved by strategically applying digital communities as vehicles for gaining influence in political and industrial domains, which has accelerated the public visibility of SWFM.

Digitally organised communities

In recent years, the development of digital platforms has provided organisational infrastructures for new forms of social movements. Social media sites now function as core structures of pro-environmental ideas and actions (Earl and Kimport 2011). Referring to this development, Lance W. Bennet and Alexandra Segerberg suggest that large-scale, fluid networks have emerged in late modern societies and that formal organisations are losing their grip on people: 'These networks can operate importantly through the organizational processes of social media, and their logic does not require strong organizational control or the symbolic construction of a united "we"'. (Bennet and Segerberg 2012, 748). Jeffrey Juris argues that digitally based social movements are 'rhizomatic' in the sense that they are constantly emerging and fusing together (Juris 2005, 199). In our analysis of the SWFM, we tap into the discussion of digitally organised networks and ask how an organisation of such a fluid constitution even evolves. Furthermore, this offers us an opportunity to look into the way that the 'united we' and the concept of community have achieved new meanings and usages in the last decade.

The literature on community suggests that the formation of community hinges on 'difference': it revolves around the identity of a social group as opposed to other groups. Rather than challenging the validity of this perspective, we explore how the study of recent digitally organised pro-environmental communities contains a critical element. Instead of being defined first and foremost in relation to other contemporary entities, such communities are loosely composed around a shared sense of *urgency*: a desire to bring about particular futures – in our case, a future without food waste. Therefore, the relationship that defines the community is not synchronic – i.e. constituted through the opposition to other contemporary communities – but rather diachronic: it is stretched in time towards potential future outcomes and consequences. We suggest that with the advent of social media, communities are more than anything assemblages of potentials that are perpetually on the brink of being realised. We contend that attention to the future poses fundamental questions about the way that the concept of community has often been applied in anthropology and related disciplines.

The Stop Wasting Food Movement

The corner café in the city centre of Copenhagen is packed with people. In the middle of the room there is a large bar with beers on tap and all sorts of spirits standing on glass shelves. The grey walls, in conjunction with the large chandeliers, create a distinct coarseness. I (Lex) am waiting for Selina Juul, the charismatic founder of the Danish SWFM, who has recently rocketed to fame, achieving the title 'Danish Citizen of the Year' in 2014. While waiting, I browse the SWFM website on my Smartphone, which says that it is a Danish non-profit organisation founded in 2008. The website says that Juul works towards a continuous reduction of food waste, primarily in Denmark. It also says that the movement has helped to bring about a massive focus on food waste. With the support of over 60,000 people, including a number of prominent politicians and acknowledged food personalities, it has achieved a large number of positive results.

A woman interrupts my Internet browsing. From photographs I have seen and a recent seminar that I attended, I immediately recognise her as Selina Juul. She takes off her black sunglasses and gives me a hug. She apologises for being late and suggests that we go to the café's second floor to talk about her movement.

Amid the hustle and bustle of the café on a Tuesday afternoon, Selina Juul introduces the development of her movement over the last 10 years. She describes how she came to Denmark as a Russian emigrant. From her experience growing up in a place with scarce resources, she was outraged by all the food waste in supermarkets and private households in Denmark. She explains:

> I was an emigrant from Russia, and I was totally shocked that you discard so much food in Denmark…. When communism collapsed, there was no food in the stores, and when I came to Denmark there was plenty of food in stores and a lot of it was wasted … so basically, I started [SWFM] because I was angry, and I felt that we had to do something about all the food waste. Apparently, this was a good story and two weeks later we were in the national news.
>
> (Juul, interview, 25 April 2017)

Juul adds that since she initiated the SWFM, each Danish citizen has reduced his or her food waste. She refers to a report that concludes that Danish citizens diminished their food waste by 8 per cent between 2011 and 2017. This equals a total reduction of 14,000 tonnes a year (see Miljøstyrelsen 2018). She emphasises that there is still a lot of work to do and that food waste from private households and the retail industry still corresponds to 9730 shopping carts of food every day – which amounts to 13.5 billion Danish kroner a year. She also says that the movement is situated near Kongens Nytorv, in the heart of Copenhagen, and is organised around herself, as front person, and an advisory board. She says that her work in the SWFM requires a lot of time and dedication:

I run from meeting to meeting, from event to event, from conference to conference. I answer daily 300+ mails. I participate in 2–3 interviews a week in the Danish press, and I do a lot of project management. I have also almost every week meetings with leading politicians and top managers in large corporations.

(Juul, interview, 25 April 2017)

She also says that she has developed a flexible and open organisation without one physical centre or one ideological conviction. It is important to Juul that the SWFM is practised among volunteers and organisations in a range of activities and events in different parts of Denmark. The objective of the movement is not to mobilise fixed members in a settled association, nor has it been about motivating people to follow certain values or political ideals. Instead, Juul stresses the importance of the common cause: to ensure a full stop of food waste in the coming years:

It is really quite simple. The movement promotes a reduction of food waste, and we want as many people as possible to do something about the problem.... To me it is pivotal to arrange events and actions. Nowadays, a lot of people do not only want to hear about global climate change, they want to do something about the situation themselves in order to ensure a more sustainable future.

(Juul, interview, 25 April 2017)

Juul emphasises that her movement should be able to act and communicate intelligently; in order to avoid too many bureaucratic discussions or 'non-decisions', she has established an open and agile organisation:

I have tried to develop an agile decision-making process without contradicting ideas and personal power struggles. We do not aim for a formalisation and idealisation of the movement ... SWFM is action-oriented, and I believe this is the best way to reach our objective ... I do not see our flexible structure as a weakness. On the contrary, it gives us the opportunity to integrate people and institutions with different political beliefs and positions. We develop events and campaigns where people and companies can work constructively to stop the food waste.

(Juul, interview, 25 April 2017)

Thus, according to Juul, the SWFM is an open and action-oriented movement that finds support among a wide range of public and private organisations, political actors from diverse parties, and volunteers from all parts of the country.

Building partnerships

Juul highlights how she has built partnerships with private and public institutions over the years. Since starting her movement, she has been involved in more than 200 initiatives, projects and partnerships in Denmark and abroad. This includes 'stop wasting food' actions in private households, public schools, and private corporations. She shares an example of a recent (2018) nationwide collaboration with commercial actors such as the supermarket chain REMA 1000, Salling Group, Denmark's largest retailer, and Gartneriet Alfred Pedersen & Søn, a tomato producer. The SWFM and its campaign partners reintroduced deformed fruit and vegetables and vegetables that are discarded by traditional grocery stores as 'too old' to sell on the market, selling them as 'second-hand tomatoes' and a 'new' product called 'food waste ketchup'. Juul says that the initiative has helped to reduce the waste of more than 75 tonnes of tomatoes.

As the founder and spearhead of the SWFM, Juul is an indefatigable advocate for her cause. She builds partnerships with both private and public organisations. She is well-connected with Denmark's celebrity community: chefs, actors, and politicians have expressed their support for the movement's campaigns and events. She writes opinion pieces in national and international newspapers to set an agenda in political and industrial spheres by influencing policy-makers and industrial leaders. She posts photographs of herself on social media taking a morning run with the Danish Prime Minister and meeting with the Danish monarchs and industrial leaders. Every instance of such publicity demonstrates that SWFM's goals are being introduced to influential people who may be able to influence legislation on the matter or make strategic or practical changes in large institutions and corporations. This kind of lobbying could be defined as not only interventional efforts for politically advancing the fight against food waste in Denmark and abroad. Rather, by successfully using the social media, Juul has managed to create a feedback mechanism. She gains access to influential politicians and celebrities, which increases her exposure on the social media, while she uses of this exposure to make herself valuable to politicians, who gain access to her network and profit from becoming affiliated with SWFM.

Juul's work stands in contrast to various anti-governmental activist movements, which, as suggested by the sociologist Alberto Melucci (1989), confronts dominant cultural norms and procedures and generates new social and political imaginaries that are continuously incorporated into daily life. These oppositional collectives present reforming ideas to existing political and socioeconomic positions, what the communication scholar John Downing (2003) calls 'radical medias'. In contrast to the anti-governmental activist movements, the SWFM enters into partnerships with Danish political and industrial leaders. The strategy of the movement and Juul as its leader is similar to what social media scholar Graham Meikle (2002) defines as 'tactical media', which purposefully avoids the oppositional approach. Juul continuously builds collaborations with institutions, citizens, and companies in Denmark for the purpose of mutual interest and gain – and to end food waste in the future.

Communicating on digital platforms

During her conversation with Lex, Juul explains that she had come directly from an interview with a journalist from Danish national television. She says that media attention has been 'crazy' after a video about the SWFM went viral one week ago. The video was shared on Facebook by the British Broadcast Corporation (BBC), and now millions of people across the globe have seen it. She smiles and adds that it is pivotal for her to share her work and visions in both the press and on social media platforms. She stresses that the SWFM has no members, no regular meetings, and no physical newsletters: the organisation's knowledge-sharing and communication is managed virtually over digital platforms such as Facebook, LinkedIn, and Twitter. As she is responsible for the SWFM's digital activities, Juul shares information, photographs, and videos about her work on a daily basis with these social media. She also posts links to what she defines as relevant news about food waste. She firmly believes that the social media, where people can show their support quickly and easily, are an indispensable way to engage and integrate people in the movement. She explains:

> I have had great success with mobilising people. Our Facebook page today has over 66,000 followers, and the LinkedIn page has over 2000 contacts. Together with my own digital media, the movement reaches over 91,000 people. So, when I share something, it reaches a lot of people.
>
> (Juul, personal conversation, 10 April 2018)

According to Juul, the digital media provides an opportunity to share knowledge and to mobilise people. The SWFM is thus not what might be understood as a conventional social movement, which generally requires formal membership and subscriptions as well as participation in meetings and other events in 'real' life situations (see Snow and Benford 1988). Such formalised engagement or structured arrangements are not a success parameter for SWFM. Rather, it is important for Juul to insert SWFM's information campaigns directly into the people´s 'private' digital sphere, which she can do via the Facebook site. She is satisfied if she can share, for example, recipes from her cookbook on how to make eatable food out of food waste, and photographs of her meeting with the US actor Will Smith. She adds that:

> I am responsible for our communication on the social media. I spend a lot of time posting pictures and videos of recent events and campaigns. I also share reports and studies from other national and international stop wasting food initiatives and results.... As part of this communication, I have tried to promote good examples of how producers, retailers, and citizens can reduce food waste. I have been constructive and concrete because it is the good stories and the good examples that make people change behaviour.
>
> (Juul, personal conversation, 10 April 2018)

According to Juul, and from observing the development of the SWFM Facebook site over a period of 24 months, it seems clear that while continuously sharing information about SWFM actions and events, Juul also shares information about other national and international initiatives that address food waste. In other words, she feeds the SWFM social Facebook site with information and photographs, which is called making 'traffic', and thus ensures on ongoing revitalisation of her movement. In this sense, the 'members' of the Facebook platform are constantly reminded about the importance of the common cause. According to Juul, the activities on the digital platforms are to inspire people to reflect on their daily food waste and perhaps even make them change their behaviour in a positive direction:

> We use the digital media as a channel through which we can 'push' information to our supporters. And they can easily respond with a 'like' or a comment. So, over distance and at any time we create a dialogue with people … I think this kind of digital conversation will lead to less food waste.
>
> (Juul, personal conversation, 10 April 2018)

Juul stresses that she has established a digital space where citizens and organisations can share good practices, know-how, or advice about the common cause, and she hopes that these 'posts' can serve as inspiration for people to diminish their daily food waste. She thus tries to use the digital 'channel' as a generator for making people reflect upon food waste and perhaps even change their daily consumption patterns. This digital sharing is expected to function as a form of inspirational guideline for changing people´s everyday practices.

When exploring the SWFM as a digital movement, it is important to emphasise that the platform or infrastructure where people and organisations communicate and interact is not neutral. The social media platforms – for instance, Facebook – have particular interests when facilitating the conversations on, in this case, the SWFM site. As a business, Facebook runs its platform with commercial interests and algorithms. Thus, the SWFM as a digitally structured community depends fully on the social media infrastructures and cannot act independently from managerial decisions made by a third commercial party.

We suggest that another challenging situation appears when people or organisations with certain political or economic interests manipulate the conversations and information sharing on the digital site. When infiltrating the community, such 'trolls' can rather easily steer the conversation in certain directions. This means that the digital movement is vulnerable to opposing political and commercial interests. In addition, administrators, who run the digital site, can buy 'likes' from artificial followers, giving them an opportunity to inflate the size and strength of the movement with fake profiles. These challenging conditions, we believe, are important for people to relate to when using social media as organisational infrastructures.

A community of potentials

We have outlined the development of the SWFM in the discussion above. We have suggested that the movement is founded first on ongoing projects and partnerships with institutions, citizens and companies in Denmark, and, second, on a virtual network where information about food waste is read, shared, and discussed by a number of more or less engaged people and organisations. With this description of the SWFM as the empirical backbone, we now turn our attention to an analysis of the concept of community. As mentioned in the introduction, this is relevant in order to discuss how fluid digitally-based movements operate and develop if they do not, as argued by Bennet and Segerberg, require the symbolic construction of a united 'we' (Bennet and Segerberg 2012, 748).

Besides being a term widely appropriated in daily speech, 'community' has achieved a central position in anthropological debates (Hamilton 1985, 7). While the term is commonly seen to index a shared sense of identity and belonging, a particular concern in such debates is whether community exists beyond instrumental expressions, i.e. whether community is constituted by actual cohorts of people tied together through a shared identity ('we') or (which often seems to be the case) whether the community is more than anything a concept evoked for a political end. Is it not the case that what appear at first to be integrated social wholes, prove in many cases to be made up of individuals with different agendas – or even of people who resist being sequestered under the canopy of the community? Anthony Cohen (1985, 13) argues that in many cases it is both: the community relies on the consciousness of its members, and in this sense is more than a political, oratorical abstraction.

Scholars often argue that the community is defined in relation to its *boundaries*, which encapsulate the identity of the group (cf. Barth 1969). The boundaries are important, argues Anthony Cohen, since communities are constituted relationally vis-à-vis other communities from which they are or wish to be distinguished (Cohen 1985). This idea extends (at least) back to Karl Marx and his reflections on the proletariat becoming self-conscious as a political subject. This requires, Marx writes, that people 'live under economic conditions of existence that separate their mode of life, their interests and their culture from those of other classes, and put them in hostile opposition to the latter' (Marx quoted in Katz 1992, 63). Through the active pursuit of its own, collective interest, i.e. in the struggle, people discover themselves as classes (Thomson 1978, 149). Thereby, in Marxist thinking, class consciousness – i.e. a class's own awareness of itself and its role in the historical process – is not an origin, but an achievement: it must be won (Lukács (1967 [1923]).

But not all people form a class. This is, Marx argues, the case for smallholding peasants since the 'identity of their interests begets no community, no national bond and no political organization among them' (Marx quoted in Katz 1992, 63). One could argue that due to their local interconnectedness, smallholding peasants do not hold the potential, or so Marx would claim, to develop into a *class unto itself*. Thus, from Marx onwards, the community has been considered political in the sense that it implies a relational difference: 'The word [...] expresses a relational idea: the opposition of one community to others or to other social entities' (Cohen 1985, 12). But at the same time the community is constructed around complex sets of symbols whose meanings vary among its members. This, Cohen argues, is the very reason that communities are able to integrate a variety of people under the same label of identity. This is also the case – and perhaps especially so – in relation to the 'deep, horizontal comradeship' of the larger, national communities famously explored by Benedict Anderson (1991 [1983]). Anderson depicts the heterogeneous community within the nation-state where the members do not know each other 'yet in the minds of each lives the image of their communion' (Anderson 1991 [1983], 6).

Anderson and Cohen (and Marx) show two important aspects of the way that the community is widely portrayed: the community (imaginary or not) is dialectically constituted internally through shared narratives and symbols that are reinforced in relation to other, external, contemporary communities or entities. Thus, identity and relationality are core prisms for understanding the concept. The trouble with community, one could argue, is that this is not always how the term community is used (Amit and Rapport 2002). If, as Cohen argues following Wittgenstein, a concept should be understood not by seeking its lexical meaning, but by identifying its *use* (Cohen 1985, 12), it seems significant to point out that the community in many cases becomes a label evoked mainly by political agents – not by the people who supposedly constitute the community. Thus, what are we to make of the concept of community when it emerges in contexts where its 'members' or 'followers' are unaware of their supposed affinity?

Furthermore, it appears that the use of community in relation to the case that we lay out in this chapter is not constituted first and foremost in relation to other communities. Rather, it revolves around *potentials* – both regarding the transition to a more sustainable society and in relation to the community itself, which exists *in embryo*. This community also exists at a rudimentary stage where its 'members', rather than being tied together via a shared identity, make up the early stage of what will become a future 'food intelligent' humanity, that is, people who are intensely aware of their patterns of food consumption and who act accordingly. With this short account of the concept of community, we now discuss how the SWFM revolves on digital infrastructures around a shared and binding potential, that is, an end to food waste and thus a more sustainable future.

Levels of engagement

The members of SWFM are primarily connected because they apparently have a common interest in the fight against food waste. Interestingly, it seems that the lack of organisational, ideological, or practical conformity makes the movement adaptive to new and divergent trends and partners. As Juul explains, she has, for instance, worked for the past 10 years with three different governments. She has co-written opinion pieces in national and international media with ministers from opposing political parties. Furthermore, she campaigns across industrial interests by involving competitive retailers in projects and events in order to advance the reduction of food waste in supermarkets. This, we believe, can be compared to what Bennett (2003, 154) defines as an 'ideologically thin' movement, which he does not consider a weakness because it allows for the integration of diverse political and economic positions and perspectives. Bennett suggests that such flexibility and diversity would be intolerable in centralised and ideological motivated movements, and that the ideological 'thinness' might help to avoid organisational conflicts and controversy. Juris makes a similar point: 'The absence of organisational centres within distributed networks makes them extremely adaptive, allowing activists to simply route around nodes that are no longer useful' (Juris 2005, 197).

Furthermore, the loose constitution of the digital infrastructure that we convey opens up the possibility of diverse levels of engagement. The reason for being part of the network and the level of participation in digital conversations is flexible, depending on the motivation and interest of the individual person or institution. For example, while promoted in the press by experts and enthusiasts, the SWFM also involves volunteers and less engaged people and institutions. Juul explains:

> Our supporters and followers can perform on different levels. They can be active stop wasting food enthusiasts. They can also be less active information recipients. Our movement is flexible and involves all people and organisations who wants to do something about food waste.
>
> (Juul, personal conversation, 10 April 2018)

By being closely connected to a range of individual lifestyles and private spheres, the SWFM, we suggest, facilitates diversified identifications of the common cause. Furthermore, when using the public SWFM site, people not only share comments, beliefs and actions with other 'followers'; rather, their activities are also likely to be shared with friends and family. In this sense, the actions on digital sites are performed in spaces that intertwine common public and individual private spaces. Likewise, Manuel Castells suggests that such networks have proven to be powerful engines for advancing and promoting ideas as they appear to be able to integrate the collective scope with individual preferences (Castells 2000).

This relates to sociological discussions in recent decades where scholars have argued persistently that disruption, fragmentation, and disintegration of larger coherent communities have become core principles in post-industrial societies (Putnam 2000; Macnaghten 2003). In this view, post-industrial individualisation engenders a situation where people engage with environmental politics as an expression of private hopes, concerns, and lifestyles (Bennett and Segerberg 2012). As the economist Michal Jacobs (1999) has suggested, contemporary societies have left the grand political narratives behind because of the rise of new information technologies and growing individualisation and globalisation. This has arguably been repeated to the point where it has itself become an over-arching narrative of its own: the time of the community has ended and, accordingly, 'the ideas and mechanisms for organizing action become more personalized than in cases where action is organized on the basis of social group identity, membership, or ideology' (Bennett and Segerberg 2012, 174).

Interestingly, it seems that the disruptions of the grand environmental narrative do not mean that pro-environmental organisations or movements have experienced more adversity in their efforts of mobilising people, ideas, and practices. Social movements worldwide have still been successful in assembling people and obtaining political influence in local and national political domains, as seen, for example, in Denmark with the political party Alternativet, and in Spain with the Barcelona en Comú citizen platform. In spite of this, Phil Macnaghten argues that the maintaining of one dominant environmental narrative in political spheres has produced a set of disjoined ideals or vision which struggle to make sense for everyday people, which makes it difficult to reach the necessary deep and extensive pro-environmental impact in contemporary societies (Macnaghten 2003).

The potential of virtual communities

In the digitally organised SWFM, followers apparently do not define themselves as part of a larger shared community and essentially do not interact in common physical spaces. Thus, in this case, it seems that the communality is not framed by people positioned in the digital movement; rather, it becomes a label evoked by a leading actor – the founder and front person, which for SWFM is Selina Juul. This does not mean that the activities on social media are unimportant or detached from the physical world. For example, on the Facebook platform, Juul shares advice about how to reduce food waste; as noted, this can inspire people to change their behaviour. Moreover, the digital information sharing can mobilise people to participate in specific physical events or campaigns or to do something similar in their neighbourhood. Juul explains:

> On the social media we reach people in their private home. We know that this is digital, but we also know that people are sitting and following the SWFM behind the screen in their living room or kitchen. Yes, this is digital communication, but we reach people in their private physical environment.
>
> (Juul, personal conversation, 10 April 2018)

In this sense, the digital site is not detached from actions in real life; by this we convey real life, i.e. that the SWFM through involved actors is practised in intertwined physical and digital spaces in service of a common cause. Other scholars reaffirm this by arguing that virtual, digital networks effectively can establish 'real time' interactions, communication, and information-sharing across borders and thus conveniently promote both local connectedness and global awareness (Bennett 2003; Castells 2000; Escobar 2004). Yet, in relation to our discussion, these actions do not seem to evoke a shared sense of community by the involved followers on the digital SWFM Facebook site.

A key concern of the SWFM, then, is to make people aware of their interconnectedness by providing them with knowledge about, for instance, the inherent fragility of ecosystems and the potentially catastrophic future scenario of too much food waste. It is, however, important to argue that the term 'virtual' can be applied in a double sense. It may refer to what is today commonly tied to the realm of computer technologies and IT systems, 'virtual reality'. But the term also has a history in the field of philosophy. Borrowing the distinction between the 'virtual' and the 'actual' from his predecessor Henri Bergson, Gilles Deleuze (1988) argues that while a domain of reality is characterised by its actuality in the sense that it can be empirically experienced, there is another, complementary domain which is not physically present. But this does not make this *virtual* domain less real. The virtual is substantially different from the actual (Mikkelsen 2018; Mikkelsen and Søgaard 2015), but we believe that virtual activities are no less real or more superficial than physical events and actions.

In an argument that adds an interesting nuance to our discussion, Bruce Kapferer (2005, 47) sees the virtual as a space that allows 'for all kinds of potentialities of human experience to take shape and form'. In philosophy, the concept of potentiality refers to an inherent ambiguity or what Agamben (1999, 179) calls 'the existence of non-being, the presence of an absence'. While the potentiality is a capacity to act in a certain way or to cause particular effects, Agamben argues (following Aristotle) that what differentiates potentiality from other forms of being is that the potential to happen is also the potential not to happen: all potentiality is also *impotentiality*. Potentiality can be conceptualised as the possible actualisation of a hidden capacity (Bateson 2000, 401). This means that a present force, figure, or formation is experienced as encompassing a source for becoming other than what is immediately present (Vigh 2011).

If this is so, then what does the potential of a virtual or digital community mean? Due to the inherent loose structural formation of the digital community (for instance, it may be constituted largely by 'likes'), the degree of devotion behind these likes is opaque and does *not* point to a social formation that may be mobilised. Using a Deleuzean term, one might say that the potentiality assumes a virtual mode of existence, which, however, does not make it 'unreal' (see Kapferer 2005). In other words, to understand movements such as SWFM, which are based on social media platforms, requires us to locate a 'virtual' – in both senses of the word – made up of possible becomings. Through an effective wielding of the possibilities offered by the social media, SWFM has attempted to unlock the present, allowing its followers to imagine the potentials that lie dormant and offering a new hope rather than succumbing to feelings of helplessness and even despair. The SWFM is not tied together via a shared identity and offers no radical antagonistic challenge to the current system. Instead, Selina Juul has built a virtual community that revolves around a potential: an end to food waste and thus more sustainable future societies.

Strong use of weak ties

As mentioned, the SWFM is continuously cultivated and organised by digital infrastructures on social media platforms. To a lesser extent, it unfolds in physical face-to-face meetings in shared public and private spaces. Inspired by the sociologist Mark Granovetter's (1973) terminology and our empirical studies, we suggest that the SWFM is constituted by 'weak ties' and unfolds between loosely connected individuals and organisations that share varying degrees of interest in a specific cause. According to such indicators, the SWFM flourishes as a loosely tied network where a large part of the involved actors – citizens, politicians, and commercial leaders – support an overarching objective. Regardless of the level of involvement of its 'followers', the activity on the Facebook site becomes proof of an extensive community of devoted pro-environmentalists fighting a particular cause. Juul thereby enters into political, industrial and academic domains with her many followers and good (powerful and recognised) friends, giving her and her movement symbolic capital, and thus a stronger position from which to advance the Stop Wasting Food Movement.

When Juul accesses industrial or political leaders, she positions her visions at an executive level, where she has an opportunity to affect, for example, retail management or national legislation. In this way, the founding actor of the SWFM uses her digital community, in this case a large number of people, including powerful and recognised actors, as an 'inactive muscle' in order to gain influence and to set an agenda. This form of environmental lobbyism must be seen as a potential vehicle for making a constructive impact approaching the overarching aim.

Furthermore, we suggest that such 'strong use of weak ties' includes the followers of the movement. When collaborating with Selina Juul, the minister, the industrial leader, and the food celebrity use the 'brand' value of the movement. In their commitment, they tap into SWFM's symbolic value, which Juul has built over the last 10 years. For example, retailers need Juul as a partner to help them build the credibility of their efforts to minimise food waste in their supermarkets. Along these lines, by interacting with the movement, the involved partners gain access to a strong brand and a large group of people, and thus partake in the positive story of the SWFM. This 'strong use of weak ties', we believe, leads to a continuous reciprocal accumulation of symbolic capital, where the involved partners gain value by joining together in the fight against food waste.

Conclusion

In this chapter, we have discussed the SWFM as a community of potentialities. We have described how the movement is founded on trans-sectorial partnerships with institutions, citizens, and companies in Denmark. Furthermore, we have suggested that the work in SWFM is organised on a digital platform involving more or less engaged people and organisations. In this, we have highlighted SWFM founder Selina Juul as fundamental in all its activities towards the overarching pro-environmental objective.

Based on our analysis, we suggest, as argued by Cohen, that the communality in SWFM is evoked primarily by certain influential actors – in this case, Selina Juul – not by the people who supposedly constitute the community. Furthermore, we propose that the use of community in the case of the SWFM is not established in relation to other communities. Rather, we believe that the potentiality of a certain desired future, a sustainable life without food waste, functions as a binding determinant.

Finally, we note that the SWFM is very focused on the here and now – and sees the community as existing well in the future. In this sense, one could argue, SWFM is a messianism in reverse. The traditional messianic community congregates around the expectation that a future event will occur that will bring about redemption. The SWFM movement is equally optimistic, but its positive attitude is grounded on redemptive acts in the present that will lead to a grand utopian community of 'food intelligent' supporters. The successful movement thus shows that another future and other worlds are always inscribed in the present. The ongoing success of the movement relies heavily on Juul's ability to remain visible as a charismatic, public figure. However, while Juul may indeed appear to strongly resemble the charismatic ideal type imagined by Max Weber (1958 [1946]), there are significant differences. Charismatic leadership, according to Weber, is transitory 'because pure charisma does not know any "legitimacy" other than that flowing from personal strength, that is, one which is constantly being proved' (Weber 1958 [1946], 248). Juul's legitimacy is continuously affirmed by the reality that we are all, according to most forecasts, approaching a world in which ecological disasters will increasingly come to affect the lives of human beings. Thus, the project advanced by SWFM persistently makes claims to an emerging reality, which is only becoming increasingly more real and is closely interwoven with a host of projections that not only scientists and activists subscribe to: the notion that *something must be done*, is, arguably, trickling into the common awareness. With all the changes the world is currently undergoing, it appears that Juul's movement may be strengthened. This means that the followers of the SWFM are not interconnected via a shared identity or common activities. Instead, they are united by Juul around a future potential: an end to food waste.

Note

1 The distinction between the 'objective' and the 'subjective' is borrowed from Slavoj Žižek (2008). He distinguishes between 'subjective violence', i.e. violence that is inflicted by a clearly identifiable agent of action, and 'objective violence', which refers to the 'often catastrophic consequences of the smooth functioning of our economic and political systems'. In other words, this is systemic violence with no clear perpetrator, such as socioeconomic inequality. Žižek adds that systemic violence tends be invisible because it is the 'normal' state of affairs, the background against which we perceive subjective violence as disturbing (Žižek 2008, 2).

References

Agamben, G. (ed.) (1999) *Potentialities: Collected Essays in Philosophy*. Stanford, CA: Stanford University Press.

Amit, V. and Rapport, N. (2002) *The Trouble With Community: Anthropological Reflection on Movement, Identity and Collectivity*. London: Pluto Press.

Anderson, B (1991[1983]) *Imagined Communities: Reflections on the Origin and Spread of Nationalism*. New York: Verso.

Barth, F. (1969) *Ethnic Groups and Boundaries: The Social Organization of Culture Difference*. London: George Allen & Unwin.

Bateson, G. (2000) *Steps to an Ecology of Mind*. Chicago: University of Chicago Press.

Bennett, L. and Segerberg, A. (2012) The logic of connective action. *Information, Communication & Society*, 15(5), 739–768.

Bennett, W.L. (2003) Communicating global activism. *Information, Communication & Society*, 6(2), 143–168.

Castells, M. (2000) *The Network Society*, 2nd edn. Oxford, UK: Blackwell.

Cohen, A. (1985) *The Symbolic Construction of Community*. London: Routledge.

Deleuze, G. (1988) *Bergsonianism*. New York: Zone.

Downing, J.D.H. (2003) The Independent Media Center movement. In N. Couldry and J. Curran (eds) *Contesting Media Power*. Lanham, MD: Rowman & Littlefield.

Earl, J. and Kimport, K. (2011) *Digitally Enabled Social Change: Online and Offline Activism in the Age of the Internet*. Cambridge, MA: MIT Press.

Escobar, A. (2004) Actors, networks, and new knowledge producers. In B. de Suosa Santos (ed.), *Para Além das Guerras da Ciência*. Porto, Portugal: Afrontamento.

Granovetter, M. (1973) The strength of weak ties. *American Journal of Sociology*, 78, 1360–1380.

Hamilton, P. (1985) Editor's Foreword. In A. Cohen (ed.), *The Symbolic Construction of Community*. London: Routledge.

Jacobs, M. (1999) *Environmental Modernisation: The New Labour Agenda*. London: The Fabian Society.

Juris, J.S. (2005) The new digital media and activist networking within anti– corporate globalization movements. *The Annals of the American Academy of Political and Social Science*, 597, 189–208.

Kapferer, B. (2005) Ritual dynamics and virtual practice: beyond representation and meaning. *Social Analysis*, 18(2), 33–54.

Katz, C.J. (1992) Marx on the peasantry: class in itself or class in struggle. *The Review of Politics*, 54(1), 50–71.

Lukács, G. (1967 [1923]) *History and Class Consciousness*. London: Merlin Press.

Macnaghten, P. (2003) Embodying the environment in everyday life practices. *Sociological Review*, 51(1), 62–84.

Meikle, G. (2002) *Future Active: Media Activism and the Internet*. New York: Routledge.

Melucci, A. (1989) *Nomads of the Present*. Philadelphia: Temple University Press.

Mikkelsen, H.H. (2017) Never too late for pleasure. *American Ethnologist*, 44(4), 646–656.

Mikkelsen, H.H. (2018) *Cutting Cosmos: Masculinity and Spectacular Events among the Bugkalot*. New York: Berghahn.

Mikkelsen, H.H. and Søgaard, T.F. (2015) Violent potentials: exploring the intersection of violence and masculinity among the Bugkalot. *Norma*, 10(3–4), 281–294.

Miljøstyrelsen (2018) Kortlægning af sammensætningen af dagrenovation og kildesorteret organisk affald fra husholdninger. København: Miljø- og Fødevareministeriet.

Putnam, R. (2000) *Bowling Alone: The Collapse and Revival of American Community*. New York: Simon & Schuster.

Snow, D.A. and Benford, R.D. (1988) Ideology, frame resonance, and participant mobilization. *International Social Movement Research*, 1, 197–217.

Thaler, R.H. and Sunstein, C.R. (2008) *Nudge: Improving Decisions about Health, Wealth, and Happiness*. New Haven, CT: Yale University Press.

Thomson, E.P. (1978) Eighteenth century English society: class struggle without class? *Social History*, 3, 146–150.

Vigh, H. (2011) Vigilance: on conflict, social invisibility, and negative potentiality. *Social Analysis*, 55(3), 93–114.

Weber, M. (1958 [1946]) *From Max Weber*. New York: Oxford University Press.

Žižek, S. (2008) *Violence*. New York: Picador.

9 There was no 'there' there any more

An Australian story about knowledge, power, and resistance

Thomas Hylland Eriksen

Introduction

The decade of the 2010s has given us concepts such as 'fake news' and 'alternative facts'. It has seen the proliferation of conspiracy theories, mistrust in scientific research, partial accounts parading as 'the real truth which has been concealed from us, the people', and revolts against allegedly smug academic elites and distant political elites. YouTube videos claiming that research on climate change are a scam get far more views than videos presenting the science of climate change. In this world, where the authority of science and empirical methods are being questioned and where even world leaders may brush aside uncomfortable facts as 'fake news', it is increasingly difficult to know whose knowledge to trust and how to act upon trusted knowledge in situations where something important, such as people's livelihoods, is at stake.

The acceleration and intensification of global processes has led to overheating across the world in the sense that change now takes place faster and with more wide-ranging consequences than before (Eriksen 2016a). Globalisation in its twenty-first century incarnation is a complex and uneven series of processes marked by crises that are often perceived as being ultimately global, but which remain local in their effects. Economic downturns, inequalities, and alienation resulting from large-scale corporate capitalism, environmental destruction, and climate change are all familiar sources of destabilisation in our day and age, with the nexus of knowledge and power – contested, changing, but often hegemonic – being a privileged site for the exploration of the crises of globalisation and the conditions for establishing alternatives.

In order to study the particular sociocultural configurations that emerge in response to fast, typically exogenous change, my recent research in Gladstone, Queensland (Eriksen 2018) has led me to study the relationship between knowledge and power in some detail. This topic, and contestations between different knowledge regimes, would become a far more important subject than I had anticipated, and it is one that is important to the residents of Gladstone. In this research (Eriksen 2016b, 2016c; see also Eriksen and Schober 2017), I ask how different kinds of knowledge are being articulated with each other in situations of social or cultural transformation and friction, to what extent and in what ways one form of knowledge becomes hegemonic and politically decisive, and what the conditions are for alternative modes of knowledge as the basis for resistance or alternative courses of action.

Knowledge in anthropology

Whether it is planned or unplanned, rapid change has unintended side effects, is understood differently by people in different subject-positions and tends to be contested by those who are immediately affected negatively by the changes. The ethnographic case that I discuss here is marked by great internal heterogeneity when it comes to making sense of change: actors and stakeholders not only respond in different ways, but they frequently describe facts differently as well. In connection with infrastructural projects, from mining to industry, from road construction to real estate development, investors, politicians, media, non-governmental organisations (NGOs), and locals directly affected perceive these processes and their implications differently, drawing on different sources of knowledge and representing different interests and agendas. Frequently, 'expert knowledge' is contrasted with 'experience-based knowledge', but different kinds of disembedded expert knowledge may also clash, as when independent researchers reach results that are at odds with reports commissioned by industry or government. The contrast between embodied and cognitive knowledge has always been important in anthropological research on knowledge regimes and their relationship to social action.

An exceptionally rich and fertile field of research and theorising, the study of knowledge systems has for many years raised epistemological, methodological, and indeed ontological questions within the anthropological discourse about cultural diversity. The great rationality debate following the philosopher Peter Winch's (1964) critique of E.E. Evans-Pritchard's analysis of Azande knowledge about witches, summed up in the latter's assumption that witches do not really exist (Evans-Pritchard 1983 [1937]), comes to mind here. (The discussion itself was an indirect descendant of the controversy concerning the Sapir-Whorf hypothesis on the linguistic construction of reality.) Questions concerning translation, commensurability, hegemonic knowledge, and ethnocentric bias were taken up and rephrased recently by Viveiros de Castro (2004) and his followers, who go beyond theorising about knowledge and rationality by arguing that worlds inhabited by humans may be radically different 'all the way down'. However, in the ethnography from industrial Australia that I present here, the relevant aspects of knowledge systems and regimes can be studied, understood, and compared by using the conventional methods of anthropological fieldwork, interpretation, translation, and comparison. There is no incommensurability. The multiple, often converging crises of globalisation, I argue, are best addressed by understanding how knowledge constructions relate to power and change and how knowledge regimes articulate with each other.

As regards the contrast between cognitive and embodied knowledge, the Greek concept of *habitus* was most famously developed in contemporary social theory by Pierre Bourdieu (1977 [1972]), who in his theory of practice sought to come to terms with power as a multidimensional phenomenon expressed through symbolic and cultural struggles even if it was constituted in politics and the economy. Habitus, a term with origins in Aristotle's philosophy, was the connecting point between individual actors and the larger system, a form of internalised knowledge situated in the body that signals the implicit and non-verbal rules of a particular configuration. A close relative of Paul Connerton's (1989) concept of habit-memory, itself inspired by Maurice Halbwachs' (1950) Durkheimian sociology of social memory, habitus, or tacit, embodied knowledge has represented a methodological challenge to anthropologists: it is understood by doing, not by talking (see Hastrup and Hervik 1994). I shall not explicitly address the issue of how knowledge becomes embodied; instead, I raise questions about the relationship between different kinds of knowledge regimes and their respective relationship to power. In fact, Bourdieu's distinction between *doxa* and *opinion* might be more useful for the task at hand than his concept of habitus, doxa being the implicitly held beliefs that are usually not verbalised but simply taken for granted. Doxa is thus unquestioned, while opinion is recognised as being open to disagreement and therefore points towards the possibility of collective action and change.

A final family of approaches that needs to be mentioned in this context is that associated with Edward Said's *Orientalism* (Said 1978) and postcolonial theory, Michel Foucault's archaeology of knowledge (Foucault 1970), James Scott's contrasting of abstract state knowledge and concrete local knowledge (Scott 1999), as well as Bruce Kapferer's studies of ideology and state power (Kapferer 2011; Hobart and Kapferer 2012). All these bids to connect ideology, knowledge, and power are indebted to Antonio Gramsci's (1971) Marxist theory of hegemony, originally formulated when Gramsci was a prisoner under Mussolini's Fascist regime in the 1930s.

Contested knowledge

The issues faced by local people trying to make sense of global worlds may be illuminated through the concept of *clashing scales*: local, context-specific forms of knowledge frequently contradict or simply present a different version of reality to the standardised, abstract forms of knowledge that may stem from the dominant global economic system and/or the state (Eriksen 2016a). Norman Long's (1989) concept of 'the interface', introduced to account for the clashing worlds of native South Americans and development agencies, exemplifies a phenomenon of far more general significance than the single case he looked into: abstract expert knowledge usually overrules local, partially embodied knowledge. Clashing scales are also at the heart of many forms of anthropological engagement, from James Scott's (1999) study of state interventions to Claude Lévi-Strauss's (1976 [1955]) mournful lament of the loss of indigenous worlds to the benefit of a flattening modernity. Therefore, if we are to look at knowledge and power under conditions of overheating, it becomes a matter of paramount importance to understand how power is scaled and how knowledge is both transmitted along those scales and becomes entangled in the kinds of conflicts that arise in multiscalar settings.

While the current approach is informed by the aforementioned authors and many others, it is distinctive in that it emphasises the problems associated with conflicting knowledges clashing in one and the same social field, frequently leading to open disagreement, distrust, and challenges to various claims of legitimacy. When, for example, there is a perceptible gap between experience-based knowledge and expert knowledge, the decision-making process comes under scrutiny and may be questioned or deemed illegitimate by people affected. For example, in assessing the conditions for the opening of an open-cut mine, be it in Australia or elsewhere, forms of knowledge may include that of economic profitability (the corporations, the national government), jobs (local politicians), ecological consequences (environmental NGOs), and a range of local knowledges which may emphasise, for example, changes in the local quality of life, reduced access to water, increases in the cost of living, but also increased economic opportunities. There exist different, and often conflicting, interpretations of (and, accordingly, proposed courses of action) anything from economic crises, immigration, environmental issues, and political reform to electricity generation, foreign investments, and indigenous rights.

By calling attention to the relationship between knowledge and interests, local and trans-local levels of decision-making and local responses to rapid change, I invite you to explore with me the question of why certain versions of the world become hegemonic and what it is that triggers adherence to particular facts and interpretations – and, no less interesting, what makes people change their minds. The question 'Who to trust?' is fundamental, and is usually supplemented by the question 'Why should I trust them?'. This is not a foolish question animated by vitriol and misinformed by ignorance of scientific knowledge, and the answer gives a direction to possible courses of action.

I now move to my empirical case, starting with an overview, before zooming in.

Coal, gas and the Australian dream

Mining is important to the Australian economy: the sector represents about 10 per cent of the country's GDP. Only 2.2 per cent of the labour force was employed in mining in 2015, but it contributes indirectly to other sectors by generating a demand for services and auxiliary industries and through taxes and royalties to the states and the federal government. About 80 per cent of the electricity in the country is generated by coal.

Australian national identity is also to a considerable extent connected with mining. The successive Australian gold rushes from 1851 onwards brought waves of immigrants, mainly European, to the country and created fortunes locally. Rags-to-riches stories made their way into local folklore. The vast outback and desert areas that make up much of the continent tickle the collective imagination through their vast repositories of invisible wealth in the form of gold, uranium, oil, coal, and other valuable minerals. Since 1960, manufacturing has declined steadily in economic importance (Langcake 2016), while the extent and economic importance of mining have grown. In recent years, the extraction of unconventional fossil fuels (shale oil, coal seam gas) has added new sources of wealth to the existing resources.

Some of the richest coal fields in Australia are in Queensland, and much of the coal is shipped from the port of Gladstone, a small city of 40,000 inhabitants (around 60,000 if the commuting area is included), but which boasts one of the world's largest coal ports, second in Australia only to Newcastle, New South Wales. Until 1967, however, the town was mainly integrated economically with the region and had no fossil-fuel related industry. The cornerstone enterprise was Swift's Meatworks, which grew in importance and prosperity as a supplier of tinned meat (mainly spam) to Allied forces during the Second World War. The Meatworks were closed down in 1963, and on the very same site, one of the world's largest alumina refineries was opened in 1967. QAL (Queensland Alumina Ltd), which would eventually get its electricity from the new coal-driven power station on the edge of town, opened in 1976, and the alumina would be turned into aluminium at the nearby Boyne Island Smelter from 1982. With the opening of the Moura railway line for transporting coal from the interior of Queensland in 1968 and the construction of a coal terminal at Barney Point, Gladstone had in the space of a few years become a fully-fledged industrial city.

Industrial and coal-related developments in the Gladstone region have continued at an uneven pace. A second alumina refinery is located in Yarwun, west of the city; Cement Australia (formerly Queensland Cement & Lime) operates a factory at Fisherman's Landing just north of Gladstone and a mine in Mount Larcom to the north-west. There is a quarry, a chemical factory, and many auxiliary activities – scaffolding, mechanical workshops, transport companies and so on – adding to the industrial, and industrious, face presented by Gladstone to the visitor.

From the early 2000s until the decline in coal and gas prices in 2014, industrial growth accelerated in Gladstone. Beginning in 2011, three plants for the liquification and storage of coal seam gas were built on Curtis Island across a narrow strait from the city, the first liquid gas being shipped in 2014. Ground was cleared and pipelines stretched several hundred kilometres to transport the gas from the interior of the state. Simultaneously, a new coal terminal was built at Wiggins Island just north of the city, increasing the port capacity considerably. To enable access for large ships, the western, shallower parts of Gladstone harbour were dredged from 2011 to 2013, removing 36 million cubic metres of sediment in the process.

The ownership structure in large industrial operations in Gladstone is complex, and the liquid natural gas (LNG) facilities are no exception. Although the Gladstone Ports Corporation is state-owned, the projects are owned by consortiums consisting of several companies with a complex and transnational ownership structure. The LNG plants were built by the American corporation Bechtel, but the three plants are owned and operated by others. One is owned by QCLNG (Queensland Liquid Natural Gas), which in turn is owned by BG (British Gas); one is jointly owned by the Australian energy company Santos and the Malaysian company Petronas; and the third project, Australia Pacific LNG, is operated jointly by the Australian company Origin, the American company ConocoPhillips, and the Chinese company Sinopec.[1] Moreover, many subcontractors have been involved in various stages of the construction of pipelines and the plant itself, from large engineering companies to small, local actors such as a local transport company that moved trucks by boat from the mainland to the island and back. When something goes wrong, it is therefore not always easy to identify who or what to blame and what to do, given the intricacies of the logistics and the complex ownership structure.

Conflicting knowledge regimes

The main research question raised in my fieldwork (2013–2014) in Gladstone concerns local responses to these changes and the unintentional side effects of rapid industrial growth more generally (Eriksen 2015, 2016b, 2016c, 2017, 2018; Eriksen and Schober 2016; Pijpers and Eriksen 2018; Stensrud and Eriksen 2019). There is a diversity of local perspectives, but there is a broadly, but not universally, shared indignation over the large corporations' failure to engage in a direct and sustained way with the local community. Many claim that Gladstone has received far too little in return for allowing large-scale industrial developments: an environmental activist went so far as to describe the city as 'the sacrificial lamb of Queensland. If it is noisy and dirty, just put it here.' It is commonly heard that the region produces considerable wealth for south-eastern Queensland (the Brisbane and Gold Coast area) and gets little in return.

On the other hand, there was less explicit ambivalence in the local population towards the double bind between fossil fuel-powered growth and ecological sustainability than I had expected. Instead, I found a great concern with health issues related to emissions, discharges, and working conditions in the industry, as well as conflicts over the status of particular forms of knowledge. For example, during my fieldwork, there was massive disagreement over the implications of the dredging of the harbour, which was alleged to carry with it substantial environmental destruction. Fish and crab disease, dying seagrass, and disappearing dolphins had effectively forced the Gladstone fishermen out of business, but the facts, causes, and effects of this issue were vigorously disputed. Interestingly, it was not merely a matter of technical, scientific knowledge conflicting with experiential knowledge, but the scientists disagreed among themselves, as did the people basing their conclusions on their own experiences (see Eriksen 2017 for the full analysis).

Another case concerns a simmering, long-term conflict over groundwater and mining in the agricultural township of Mount Larcom just west of Gladstone. While mining executives and their scientists insisted that the large amounts of water pumped out of the limestone mine did not affect the ground-water, the farmers drew on various sources of knowledge, including scientific models, to counter this view (Eriksen 2018).

It should be noted that infrastructural developments are qualitatively different from changes at the level of culture and representations. Changes in people's notions and vocabulary can happen quickly and unnoticed; changing practices takes longer, but infrastructural developments are difficult to reverse. The moment the harbour has been drained for the sake of access for large ships, or the Central Business District (CBD) of Gladstone has been dissected by a new access road to the port, or the East End Mine is in place, reversing these changes is long-term and difficult. In this sense, Emile Durkheim's suggestion in *The Rules of Sociological Method* (Durkheim 1982 [1895]) that social facts should be treated as things – *comme des choses* – was misleading since facts change or stay the same according to different logics. When Jean-Paul Sartre, in *Critique of Dialectical Reason* (Sartre 1991 [1960]), speaks about material structures as '*le champ pratico-inert*' ('the practical–inert field'), he emphasises this difference.

The case I now describe in some detail shares many formal characteristics with the two controversies I have just mentioned but differs in that the rapid change taking place in the community in question has not led to overheating, but rather to cooling down.

Deceleration in Targinnie

At a time characterised by accelerated change, certain places, activities, and processes slow down. In the global neoliberal economy, the principle that investments, production, and transactions should be based on calculations of comparative advantage inevitably lead to the downfall and marginalisation of places that were formerly significant. A big fish in a small pond easily becomes so small as to be nearly invisible or else eaten by larger fish in a large pond. This is a direct effect of the expansion of systemic boundaries and affects economic activities that are incorporated into larger systems. Changes in transnational terms of exchange, or market fluctuations, may also render boomtowns into abandoned backwaters. In a study of the Zambian Copperbelt in the 1990s, James Ferguson (1999) depicts an urban population yearning for a modernity they have lost, reminiscing about the very period when anthropological studies of the Copperbelt depicted a region and a continent about to enter the world stage. With the more recent Chinese investments in the Copperbelt, the region may be entering a new 'boom' period, doubtless with a different set of outcomes and side effects.

Acceleration takes place at the sites which at any given time count as centres, while peripheries may become even more peripheral than they were. Such was the story of Targinnie, a rural community just north of Gladstone, east of Mount Larcom.

Settled in the late nineteenth century, Targinnie was never what would be called a bustling town, but it had a school, a community house, and, for a time, a village shop. There were somewhat successful attempts to mine gold in the area, but the economic mainstay of the community was agriculture. Until the late twentieth century, the community was integrated at a regional level. The fruit – pawpaws and mangoes were local specialties – was sold in nearby towns or tinned at the nearby Golden Circle Cannery, and meat and dairy products were likewise marketed locally. The scale was manageable, and people knew where to go and whom to address with their complaints. In the 1960s, more than 100 families in Targinnie were full-time farmers. By the mid-1970s, only 18 were left. By 2013, there were none, and the settlement was all but deserted. What had happened?

Notably, Targinnie became integrated into larger systems of production and exchange owing to improved facilities for refrigeration and more efficient transport, but also the steady replacement of small companies with large corporations. This meant, first, that the small-scale fruit producers would have to compete with the larger producers in the wetter and more fertile lands further north, which was economically difficult. Second, an improved infrastructure made mobility into Gladstone easier from the 1980s. Third, new job opportunities due to the industrialisation of Gladstone lured many away from agriculture to factory work. This in itself could explain the community's marginalisation, a situation shared with rural areas in many countries, but this kind of change would not be sufficient to depopulate the place completely. When I mentioned Targinnie to a woman in Gladstone, she exclaimed: 'Well, they were poisoned!' Others mutter, wistfully, that it was 'a shame' what happened to Targinnie. Let me explain.

The heat briefly enters the town: the oil shale adventure

Oil shale was discovered in the area as early as the late nineteenth century – ironically, the discovery was made by coincidence during an early dredging operation in the Narrows (Blake 2005, 36). There had been intermittent interest in extracting the shale oil since the 1940s, but it was only in the late 1990s that plans were realised to exploit the resource commercially. When the Stuart Shale Oil plant was built in 1998–1999, the community, having thrived on the coexistence of agriculture and industry for decades, took a generally positive view at the outset.

Southern Pacific Petroleum (SPP) and Central Pacific Minerals had explored the oil shale since the 1970s, entering into a partnership with the Canadian company Suncor Energy, which allegedly had developed appropriate technology, in the mid-1990s. The plan was to proceed in three stages: first, to build a demonstration plant, second, a slightly larger production plant, and, finally, a large factory capable of producing 85,000 barrels of oil a day. However, the investors created some problems for themselves when they started, which inadvertently helped protect the community. If they had taken out their lease under the Petroleum and Gas Act, they could have drilled anywhere. Instead, they took it out under the Mining Act, which implied that they couldn't come within 200 metres of Targinnie's pawpaw orchards. By contrast, coal seam gas leases are always regulated within the Petroleum and Gas Act, which implies that the wells can be placed anywhere.

The construction of the demonstration plant was completed in April 1999, and tests began in July. At a well-attended opening day event in August 1999, the spirit in the Targinnie community was still upbeat: most of the inhabitants saw the new factory as a welcome addition to the regional economy.

The positive attitude would not last. The nearest neighbours to the demonstration plant were bothered by the noise and lights coming from the factory, and soon a bad smell, described by an ex-resident as 'a mixture of fresh bitumen and burning tyres', began to drift across the community. In October 1999, the factory malfunctioned, sending a thick cloud of black smoke across Targinnie. One ex-resident, who lived at the time on the far side of Targinnie, remembers returning home from work in Gladstone to see his family suffering from red eyes, runny noses, headaches and nausea.

Believing in democratic procedures, some residents formed a civil society association, the Yarwun–Targinnie Representative Group. Meetings were held with local government, state government, and the factory owners. At this time, 1999, instantaneous communication was less widespread than today, and the small group came together following conversations at the village shop, during phone calls, and while volunteering. Practically everybody knew everybody in the village, and the group knew that they could rely on the support of the local community. Today, they might have formed a Facebook group.

The health complaints continued, and the air pollution became a growing concern throughout the year 2000. There were also other serious incidents resulting from malfunctioning, which occasionally enveloped much of Targinnie in smoke.

Local residents were nevertheless being told by Suncor that emissions were within acceptable limits, since the quality of the air was being monitored. In other words, measurements of the air quality (however random and incomplete they might be) were deemed more trustworthy than knowledge based on experience.

Eventually, most of the residents in Targinnie, fearing for their health, wished to leave. Some ran family farms that had been operating for 100 years. The Russian immigrants had a tightly knit community of Old Believers centred on a spartan, but functional church off the Targinnie Road; this community would now be scattered across Queensland and beyond. Most Targinnians now simply wanted to get out, fearing for their health and quality of life. Several cases of cancer had appeared. Some of the victims died, while others recovered. The residents, worried that the pollution from the oil shale plant might have triggered the cancers, were told by health authorities that the numbers were 'too small to be statistically significant'. The feeling of not being taken seriously by the authorities started to grow.

It soon turned out that the properties, located in what is arguably the most scenic and beautiful rural area near Gladstone and blessed with excellent soil and good water, could no longer be sold. Prospective buyers were aware of the emissions from the oil shale factory and the plans to expand it. This was a serious blow to the locals. Many had regarded their house and property in Targinnie as their pension fund and were planning to sell out and move south upon retirement. This investment had now shrunk, almost overnight, to nought.

Resistance through knowledge production

During the liminal period when this ultimate outcome was uncertain, the Yarwun–Targinnie Representative Group decided to fight for recognition, reparations, apologies, and compensation. At each consecutive meeting with the public relations people of the company, they were met with denial. 'They just spoke *to* us, not *with* us', says an ex-resident who was present at these meetings. In early 2002, Southern Pacific Petroleum released a mandatory, supplementary Environmental Impact Statement (EIS, Sinclair Knight Merz 2002). Along with the Yarwun–Targinnie Fruit and Vegetable Growers Association, the group contacted a legal scholar, Dr Richard Whitwell at the Department of Environmental Management, Central Queensland University, for assistance in preparing a thorough response to SPP's assurances that everything was fine and that things would only get better. Together with researchers Andrea van der Togt and Malcolm Scandrett, Whitwell produced a small book (Whitwell 2002; van der Togt 2002; Scandrett 2002) based on interpretations of the law, previous experiences in the community, questionnaire surveys, and qualitative interviews with residents. The Fruit and Vegetable Growers Association also produced a 39-page submission of their own, which was drawn upon and analysed by Whitwell and his colleagues.

The critique of the EIS is scathing and comprehensive. First, it questions the methodology used to determine the air quality and effects of emissions. Although the company states that it has monitored air emissions, the positioning and usage of the receptors is not specified. No Targinnie residents remembered having seen any of these contraptions on their property, but one reports that he had noticed a receptor which was sometimes turned on after trial runs but never during regular factory operations.

Second, the EIS refers to an 'independent toxicologist' consulted to determine the levels of dioxins in the air, but this expert is not named, and the group of residents requested the views of another toxicologist. Since values are not specified in the report, and measurements are seen to be inadequate, the results may well be doubted, Whitwell concludes.

Third, the terminology used in the report is seen as wanting and inaccurate. Whitwell points out that the term 'elevated emissions' used in the EIS is not defined. The 'bad odours' referred to in the EIS should, in his view, be replaced by 'odour/irritant'. (Some of the local residents would have preferred the term 'toxic fumes'.) Moreover, the EIS is replete with round formulations such as SPP promising to ensure that 'these issues are given close attention', without saying how. It also boasts that development of Australia's oil shale resources could lead to self-sufficiency in oil for at least 50 years, and Whitwell proposes that this view 'be reviewed and a more realistic situation outlined'.

Fourth, the experience of the residents is dismissed in the EIS as being irrelevant, which the locals perceive as being especially disturbing. The EIS refers to the 'unnecessary concerns in the community' (Whitwell 2002, 8), suggesting that they are complaining for no sound reasons. As opposed to the quantitative measurements indicating that the air quality in Targinnie was really very good, the report 'suggests that the use of human subjects is the best method for detecting the impact of odour emissions' (Whitwell 2002, 5).

Fifth, the report shows that the actions proposed by SPP are not described in a specific way, which inevitably gives the reader the impression that they are empty promises. For example, regarding dioxins, the EIS states that 'tenfold reductions are probably achievable', but Whitwell suggests that this statement 'be removed from the report or a detailed explanation with evidence on how this is to be achieved' be added (Whitwell 2002, 5).

Apart from presenting likely impacts on health, this extended commentary on the EIS also raises questions about contamination of water, the effects of airborne toxins on the orchards and livestock, and the prospects of marketing produce from Targinnie following its stigmatisation as a severely polluted area. Serious doubt is cast on the validity of nearly all aspects of the EIS, and Whitwell's report advises that the sentence claiming that 'the proponent works closely with the community' be deleted (Whitwell 2002, 6).

It was clear, then, that the people of Targinnie were not content to confront the abstract knowledge of the company with their own, experience-based knowledge, but produced their own, scientifically-based knowledge in order to get their message across, still confident that the company and the political authorities would be receptive to documented facts. The clash between knowledge regimes, in other words, did not merely contrast abstract, decontextualised scientific knowledge and the knowledge of everyday experience, but two epistemologically and methodologically equivalent knowledges. The Targinnie residents were not keen to be regarded as simpletons from the countryside.

Resisting the hillbilly narrative

Members of the community did not have the impression that Stuart Shale Oil was working with them in a genuinely collaborative way. Peter Harland, President of the Yarwun–Targinnie Representative Group, recalls that the group was

> beautifully used by them [the company] as a management tool. We were assuming that they were telling the truth when they said that this was proven technology from Canada, and that these were just small teething problems. To be honest, I took them at their word.

He would later discover that the company had more lawyers than engineers on its payroll, 'which is a bit worrying for an industrial development firm'. It soon became clear to them that the Stuart Shale Oil Plant in its first phase was not state-of-the-art technology, as they had been told, but part of the company's research and development (R&D) phase.

The farmer and local politician Craig Butler, also a member of the Representative Group, says that

> the company – SPP – and the state government wouldn't recognise that there was a problem. If there is a problem, [they ought to have said] 'Put your hand up and say yes, there is a problem, and we're going to fix it'. They accepted no responsibility, neither the company nor the state government.

During the Representative Group's meetings with the company, which took place from late 1999 to 2002, patience was wearing thin. Harland says, 'You know, I've no idea as to the number of times I've heard sentences that begin with "You have to understand …"'. He felt that they were being treated as ignorant peasants. It did not help, he adds, that the media, too, represented the Targinnie community as a collection of quaint hillbillies.

Sally, also a member of the citizens' group, is more equivocal in her judgement. Her experience of dealing with Suncor and SPP was that they 'were like any other company, they have a PR group who are nice to you, take your concerns on board, and they wanted to work with the community'. She concedes that they often came across as 'smug and superior', assuming that the Representatives knew nothing, but – taking the company's point of view – adds that it can be difficult to deal rationally with people's emotions and fears.

Personal relationships with the company representatives could be good, but no concessions were made. Suncor, which pulled out of the project in 2001, apologised orally to the Representative Group before leaving. '[The CEO of Suncor] looked me in the eye and said sorry', Peter recalls.

The Representative Group was nevertheless more resourceful than the company had expected. Craig was well connected among decision-makers and knew how corporate interests were entwined with local politics, Peter sold high-end car stereos and had a professional interest in sound and frequencies, and, as Sally, herself a knowledgeable person, says, 'Keith McGavin, yes, he's a farmer, but he's not an idiot. So yes, at the meetings, they thought that they were dealing with idiots. Then you turn up and are able to ask fairly well-informed questions'.

Michelle Butler, Craig's wife, recalls that they were being treated like ignorants by the company, and that 'people were later going into negotiations with government where they were being treated like dirt. You go on and you are just playing out a pantomime, everyone's going through the motions, but nobody seems to mean anything'.

In Craig Butler's view,

> honesty with people counts for a lot. You can say, 'I stuffed up, I'm sorry, but I'll do everything I can to try and fix it'. People will say, 'We can't ask

you any more, and they will accept it'. In this case, to put it mildly, honesty was not evident.

The Representative Group was disappointed to learn that 'the proven tech-nology from Canada' was 'a thing that fitted into a twenty-foot shipping con-tainer', as Harland puts it. They also discovered, by making their own investigations, that the plant worked well with small shale rocks and at low speed, but blew up the moment they fed larger rocks into it and increased the speed, exploding and blowing off the seals at the end of the machine. They also knew that the process could be improved, with much reduced side-effects, but that this would cost money that the company was not prepared to invest. 'They made promises to the community and the government that they would put in a catalytic burner at the end, but never did'.

The most engaged of the locals had to learn not only the language of the PR people and politicians, but also the EIS language. Some of them got to the point where they could easily find inconsistencies in EISs – which are commissioned by the companies themselves, not by the government. The companies and con-sultancy firms were not used to being questioned, since the reports were not intended to be read. 'And they don't want to know. There is a delusion factor', Peter Harland says.

> During the shale oil project, you could go down there and notice that whereas the rest of the harbour was brown from flood, yours was black and there was shimmy oil on top. I wrote a letter to the paper about this, but it wasn't printed. They were saying that the cause was fresh water flow from the Awoonga Dam [south-west of Gladstone, far from this area]. Which was rubbish.

The Representative Group soon discovered that the plant had been built 'on a shoestring', using cheap Chinese steel and no flow control in the pipelines. This meant that it was impossible to shut off valves if part of the plant malfunc-tioned; in other words, that the company was incapable of controlling its own factory.

The knowledge possessed by the local community in Targinnie, whether based on experience or on abstract concepts, was never brought to bear on the practices at the plant. Sally says that one day, people from the company would come out to their home and claimed that they couldn't smell anything. 'We said, "Seriously, can't you smell anything?" They said, "No, no, no, no"'.

Conflicts of interest

The suspicion that there were covert complicities between politicians and investors was also widespread. The political support of the project, which was 'good for Gladstone', was open and acknowledged. The Queensland Govern-ment had supported the project financially. The Gladstone City Council saw jobs, growth and dollars. Additionally, there were politicians who had invested in the plant, and who thus had personal interests in talking up the value of the shares and hushing up anything smelling of scandal or failure. Indeed, there were people in Targinnie who also bought shares in the plant, and this led to some frictions within the community.

Nor was there much local support for the Targinnie residents. 'We didn't have a lot of money or powerful contacts, unlike in Airlie Beach, where they had millions of dollars at their doorstep'. Sally is referring to a later, failed attempt by SPP to open a shale oil plant in Proserpine near the glamorous and picturesque tourist town of Airlie Beach further north on the Queensland coast. A common view in Gladstone, according to some of the people from Targinnie, was that they were 'whinging', since one is expected to put up with a bit of noise and smell if one chooses to live in an industrial area. The local newspaper, the *Gladstone Observer*, would report critically on the Targinnie emission scandal, but on the whole they expressed a trusting and positive view of the company. An ex-Targinnie resident who used to work in advertising at the *Observer*, explains that 'there is a strong incentive to defend industry too much; from an advertising point of view it is easy to see why'. In conclusion, the knowledge and experiences presented by Targinnie residents were politely received by authorities and the company, and the *Observer* occasionally published letters voicing concern. Yet it seemed as if their complaints, no matter how meticulously documented, had no effect.

To many, the experience of first negotiating unsuccessfully with the company and then being evicted and bought out at a sub-market price by the government seriously weakened their trust in the impartiality and accountability of the political system and the view that facts and knowledge mattered. 'It's money first, not people first', Sally says.

> The EIS was so flawed. There was constant low noise even for us who were not their nearest neighbours and had hills between us and the shale oil plant. The EIS claimed that the waves lapping on the shore would cover the noise, which was ridiculous, but the government accepted it.

It was when they saw the EIS, described by Michelle as 'a very lightweight EIS', that they contacted Richard Whitwell for assistance to write a proper response.

Michelle mentions a woman who worked for the EPA (Environmental Protection Agency) in Central Queensland, but who was transferred to a different district because she tended to agree with the arguments put forward by the Representative Group. 'She started to admit things that the government did not accept – they insisted that things were OK.'

Queensland Energy Resources, owners of the current shale oil plant at Targinnie, have reported that they have received no complaints from the nearby community following the announcement of their desire to expand the plant. 'No wonder', comments an ex-Targinnian, 'since there isn't one; it's all gone'.

Conclusion: follow the money, not the knowledge

When epistemic fault lines appear in an otherwise consensus-oriented indus-trial, forward-looking society, the relationship between knowledge and power is questioned, and trust suffers as a result. One ex-activist simply responded to one of my questions by saying, 'the answer is simple, just follow the money'. Many have seen their trust in the democratic character of the political system dwindle, and following the dredging scandal (which affected many) and the mine contro-versy (which affected fewer), faith in scientific results has also been weakened. Many who continue to trust scientific knowledge may now add that they no longer trust scientists until they know who pays for their research. All this cer-tainly held true for the community of Targinnie. And when the seeds of distrust are sown, they may proliferate quickly. For example, there is a widespread rumour that the routine blood testing of employees at some of the large factories in the Gladstone region, while it ostensibly concerns drugs and alcohol, is really about looking for evidence of toxins and symptoms of poisoning. Other things being equal, trustworthiness, in people and institutions, decreases with increased distance. The view shared by the Targinnie activists and others who feel overrun and treated disrespectfully is that when large-scale projects clash with small-scale realities, being right is far less important than being large. Distance also precludes close engagement with life-worlds and the knowledge of everyday experience. The scaling up of the Australian resource economy, which was criti-cised by many (e.g. Cleary 2012; Munro 2012), removes corporate account-ability further and further away from stakeholders, creating a democratic deficit and a feeling of being neglected, ignored, and overrun by the powers that be.

Yet this very distrust of established knowledge contains the seeds of possible citizen action and social change. Questioning authorities in this kind of setting can be empowering in that non-elite people propose different interpretations that lead to different courses of action. The anti-elitism witnessed in politics around Europe and North America these days is born of a related impulse, namely the conviction that people are being lied to, and that the power elites conceal important facts from them. A difference with the Targinnie case is that the outrage there was directed at big money and the complicity between gov-ernment and corporations, while the new populisms in Europe are strangely quiet when it comes to economic, and especially financial, power. Yet in both cases, the concept of 'fake news' has proliferated epidemically since it was coined in 2016 (see Frankfurt 2005 for a philosophical analysis of information beyond truth and falsity, that is, what he calls bullshit).

In the Gladstone region, the city itself and its surrounding rural and semi-urban communities have been affected by growth and change in different ways. Occasionally, parents have to collect their children early from a primary school near QAL (Queensland Alumina Ltd) because of white alumina dust blowing into the schoolyard. When you rise from an outdoor chair, you may notice that your trousers have been soiled by coal dust. These are everyday occurrences. The discovery of a colony of non-endemic fire ants, presumably from the USA, at an industrial site near Gladstone in late 2013 raised eyebrows, but nobody was deeply surprised. The population of Gladstone is accustomed to living with vulnerabilities resulting from industrial operations. When I took part in a Conservation Volunteers Australia project to clear an oceanic beach of rubbish and we came across Chinese and Korean water bottles, a Japanese Pepsi can, and an empty juice carton from Cyprus, nobody was surprised. Besides, on the horizon, we could glimpse the contours of coal ships from many countries waiting for their turn to load at Tanna Coal Terminal.

The recent, accelerated change in Gladstone – dredging, LNG plants, new coal terminal, expanding limestone mine and so on – is contingent on Australia's integration into a changing global economy, a key factor being the growing global demand for minerals. Scarcely anyone in Gladstone is against coal mining or industrialisation per se, perhaps because they are all entangled, but perhaps also because they see the difficulty of promoting a credible alternative. But many argue that local needs and community interests should be given first priority, and that this is not the case.

The two fundamental contradictions resulting from accelerated growth in a neoliberal world economy are highly visible and lead to a series of tensions and conflicts locally: the double bind between growth and sustainability and the disjuncture between small-scale community concerns and large-scale corporate interests. These are, at an abstract level, the major contradictions of contemporary globalisation. As I have shown, they are enacted, with high stakes on several sides, through an apparent competition between knowledge regimes which, upon closer scrutiny, morphs into an unequal power relationship where at the end of the day the credibility of facts and interpretations is irrelevant. Yet this very situation reveals a crisis of legitimacy which may, as in the Targinnie case, foment collective actions and demands for change.

Notably, the activists in Targinnie and in the Gladstone region generally target the distant corporations and what they see as a culture of greed, unlike many Europeans, who have fallen victim to growing scalar gaps resulting from accelerated globalisation, and blame immigrants and the political establishment instead. This difference is not uninteresting and should serve as a reminder that popular uprisings against democratic deficits and deteriorating life conditions do not necessarily result in xenophobic populism (as in European politics) or vulgar bigotry (as in the current regime in the USA), but can actually lead to grassroots activism demanding transparency, environmental responsibility, and corporate accountability. This is not a trivial insight in this day and age.

Acknowledgements

A previous version of this chapter was presented at the SUSY opening conference at the University of Copenhagen, 6–7 June 2018. I thank the organisers for their hospitality. I am also grateful to the editors for their excellent comments on the first draft. There is some overlap between the ethnographic parts of this chapter and Chapter 8 in my monograph *Boomtown* (Eriksen 2018).

Note

1 A fourth gas plant was planned but postponed indefinitely for economic reasons.

References

Blake, T. (2005) *Targinnie: The history of a Central Queensland Rural Community*. Brisbane: Queensland Government.

Bourdieu, P. (1977 [1972]) *Outline of a Theory of Practice*, trans. R. Nice. Cambridge: Cambridge University Press.

Cleary, P. (2012) *Mine-Field: The Dark Side of Australia's Resources Rush*. Collingwood, Victoria: Black.

Connerton, P. (1989) *How Societies Remember*. Cambridge: Cambridge University Press.

Durkheim, E. (1982 [1895]) *The Rules of Sociological Method*. New York: Free Press.

Eriksen, T.H. (2015) Globalization and its contradictions. In P. Stewart and A. Strathern (eds), *The Ashgate Research Companion to Anthropology*. London: Ashgate, 293–314.

Eriksen, T.H. (2016a) *Overheating: An Anthropology of Accelerated Change*. London: Pluto.

Eriksen, T.H. (2016b) Scales of environmental engagement in an industrial town: glocal perspectives from Gladstone, Queensland. *Ethnos*, 81(4), 1–15.

Eriksen, T.H. (2016c) Identifying with accelerated change: modernity embodied in Gladstone, Queensland. In T.H. Eriksen and E. Schober (eds), *Identity Destabilised: Living in an Overheated World*. London: Pluto, 77–98.

Eriksen, T.H. (2017) Conflicting regimes of knowledge about Gladstone Harbour: a drama in four acts. In T.H. Eriksen and E. Schober (eds), *Knowledge and Power in an Overheated World*. Oslo: Department of Social Anthropology, 72–97. Available at: www.uio.no/overheating.

Eriksen, T.H. (2018) *Boomtown: Runaway Globalisation on the Queensland Coast*. London: Pluto.

Eriksen, T.H. and Schober, E. (eds) (2016) *Identities Destabilised: Living in an Overheated World*. London: Pluto.

Eriksen, T.H. and Schober, E. (eds) (2017) *Knowledge and Power in an Overheated World*. Oslo: Department of Social Anthropology. Available at: www.uio.no/overheating.

Evans-Pritchard, E.E. (1983 [1937]) *Witchcraft, Magic and Oracles among the Azande*, ed. E. Gillies. Oxford: Oxford University Press.

Ferguson, J. (1999) *Expectations of Modernity: Myth and Meanings of Urban Life on the Zambian Copperbelt*. Berkeley: University of California Press.

Foucault, M. (1970) *The Order of Things: An Archaeology of the Human Senses*. London: Tavistock.

Frankfurt, H. (2005) *On Bullshit*. Princeton: Princeton University Press.

Gramsci, A. (1971) *The Prison Diaries*, ed. and trans. Q. Hoare and G.N. Smith. London: Lawrence & Wishart.

Halbwachs, M. (1950) *La mémoire collective*. Paris: PUF.

Hastrup, K. and Hervik, P. (eds) (1994) *Social Experience and Anthropological Knowledge*. London: Routledge.

Hobart, A. and Kapferer, B. (eds) (2012) *Contesting the State: The Dynamics of Resistance and Control*. Oxford: Sean Kingston.

Kapferer, B. (2011) *Legends of People, Myths of State: Violence, Intolerance and Political Culture in Sri Lanka and Australia*, rev. edn. Oxford: Berghahn.

Langcake, S. (2016) Conditions in the manufacturing sector. Canberra: Reserve Bank of Australia. *Bulletin*, June Quarter, 27–33.

Lévi-Strauss, C. (1976 [1955]) *Tristes tropiques*, trans. J. Weightman and D. Weightman. Harmondsworth: Penguin.

Long, N. (1989) *Encounters at the Interface: A Perspective in Social Discontinuities in Rural Development*. Wageningse Sociologische Studies, 27. Wageningen: Wageningen Agricultural University.

Munro, S. (2012) *Rich Land, Wasteland: How Coal is Killing Australia*. Sydney: Macmillan Australia.

Pijpers, R.J. and Eriksen, T.H. (eds) (2018) *Mining Encounters: Extractive Industries in an Overheated World*. London: Pluto.

Said, E. (1978) *Orientalism*. New York: Pantheon.

Sartre, J.P. (1991 [1960]) *Critique of Dialectical Reason*. London: Verso.

Scandrett, M. (2002) Appendix 5: Socio-Economic Report Yarwun/Targinnie Representative Group. In R. Whitwell (ed.), *Impact of Shale Oil Production on the Yarwun/Targinnie Community. Response to the Supplementary EIS*. Central Queensland University: Centre for Environmental Management.

Scott, J.C. (1999) *Seeing Like a State: How Certain Schemes to Improve the Human Condition have Failed*. New Haven, CT: Yale University Press.

Sinclair Knight Merz (2002) *Stuart Shale Oil Project, Stage 2: Environmental Impact Statement Supplementary Report*. Brisbane: Sinclair Knight Merz.

Stensrud, A.B. and Eriksen, T.H. (eds) (2019) *Climate, Capitalism and Communities*. London: Pluto.

van der Togt, A. (2002) Appendix 1: Analysis of Yarwun Targinnie Fruit & Vegetable Growers Association Inc. Responses to Stuart Oil Shale Project EIS Report. In R. Whitwell (ed.), *Impact of Shale Oil Production on the Yarwun/Targinnie Community. Response to the Supplementary EIS*. Central Queensland University: Centre for Environmental Management.

Vivieiros de Castro, E. (2004) Perspectival anthropology and the method of controlled equivocation. *Tipití: Journal of the Society for the Anthropology of Lowland South America*, 2(1), 3–22.

Whitwell, R. (2002) *Impact of Shale Oil Production on the Yarwun/Targinnie Community. Response to the Supplementary EIS*. Central Queensland University: Centre for Environmental Management.

Winch, P. (1964) Understanding a primitive society. *American Philosophical Quarterly*, 1(4), 307–324.

10 Labour organising against climate change

The case of fracking in the UK

Vivian Price

Introduction: the earth is demanding attention

Labour unions in the United Kingdom, like unions in many countries, consider the impact of climate change on their members' employment first and foremost; some unions struggle to represent broader working class interests (Räthzel and Uzzell 2011). Energy unions are at the heart of dislocation concerns but other unions represent members directly or indirectly tied to the energy industrial complex. As groups representing working people, labour unions are positioned to express demands for social justice. 'The labour movement is a potential site where a global political vision on climate change can evolve, directly confronting existing socio-ecological relations and empowering a coherent alternative vision' (Hampton 2015, 184). Labour lives within the tension between survival and solidarity.

The Trades Union Congress (TUC), the federation of the majority of unions in England and Wales, took one of its most significant positions on climate change in the Fall of 2012 in a resolution supporting a moratorium on fracking. The campaign to pass that motion and the continuing battle against fracking by activists, organisations, and unions form the crux of this chapter. Obach (2004), Mayer (2009), and Lundström *et al.* (2015) emphasise the importance of individuals, both committed unionists who work with environmentalists and individuals in environmental movements who push their organisations to cooperate with unions. Whether elected officers, leaders, or activists, these are people who broaden the perspectives of their organisations because of their familiarity with both labour and environmental cultures. This was the case with the fracking resolution: it is a story that illustrates how individuals can affect the mind-set of an organisation.

That fracking is new to the UK and there are few workers presently employed in the industry makes this struggle easier to oppose than if there were hundreds of workers that would be displaced. But, certainly, workers and unions are interested in what they see as potentially good paying jobs, and the fracking industry has a huge investment in forecasting a large workforce to incentivise support. There is a mutual interest between business and government in developing fracking, and public policy has been made more favourable to fracking than, say, to the development of offshore wind power (Graziano *et al.* 2017).

The fracking industry is determined to drill for shale gas in the UK, and their desire to do so has been adopted by the ruling Tory Party. Advocates point to the growth and apparent success of fracking in the US, but fracking has many detractors (Foran 2014; Christopherson 2015; Rousu *et al.* 2015; Short and Szolucha 2017; Clough 2018). There is a lot at stake for the companies promoting onshore gas drilling in the UK and beyond its shores: much of Europe is considering gas exploration through fracking. As of this writing in late 2018, New Preston Road in Lancashire is the first and only place in Europe where new commercial exploration for shale gas is underway. There has been fracking in the Netherlands, but a 2015 moratorium put drilling in abeyance, at least temporarily (Metze and Dodge 2016). Thus, the outcome of the fracking debate has potential significance beyond the UK (Bradshaw and Waite 2017; Short and Szolucha 2017).

Opposition to fracking is based on environmental reasons, including water pollution and waste, land destruction, methane emissions, and health and safety issues. Another key reason for opposing shale gas is that it prolongs the reliance on fossil fuel and business as usual into the future, undermining the shift to renewables and a more equitable society. Support for fracking parallels the reasons in reverse, focusing on economic and security reasons and job development, together with the claim that gas is a 'bridge' fuel towards renewables (Jaspal and Nerlich 2014; Foran 2014).

The fracking industry promises a boost in employment for the traditional blue-collar skilled trades employed in fields such as natural-gas extraction, well drilling, and pipeline transportation, but also tends to factor in an increase in employment in freight trucking and highway, street, and bridge construction (Foran 2014). In other words, the fracking supporters exaggerate the direct employment that the industry produces (Maniloff and Mastromonaco 2017). It also should be noted that drilling is short-term and often does not employ local workers.

Union response to climate change in the UK has a strong bottom-up history, and union opposition to fracking can be traced from 2012. There has been considerable community and environmentalist resistance to fracking, including resistance encampments, following the tradition of the climate camps and anti-nuclear camps of the past decades. The British labour movement provides space and opportunities for rank and file workers to raise their voices at many levels, for example, through motions that can be passed on the regional level and then introduced and voted on nationally. The union system of environmental representatives that emerged around 2006 with the Trade Union Congress's Green Workplaces project (Hampton 2015, 127) creates a space for rank and file workers to push issues forward. Union members can hold their elected officials accountable at meetings and through petitions. They also work in coalition with community groups and environmentalists in organisations such as the Greener Jobs Alliance, Campaign Against Climate Change, and Hazards, as well as Trade Unions for Energy Democracy (TUED), an international group with a solid base in the UK. There are also smaller and informal groups in locations around the UK who meet online and at protest sites.

At the same time, it is clear that some labour unions are better able to explicitly take up the mantle for climate action because the majority of their members are not employed in jobs in the fossil fuel industry. Occupational positionality – that is, how closely one's job is tied to the fossil fuel industry – tends to have a great influence on responses to climate change, in this case, on fracking. Activists and leaders espousing various viewpoints are always exerting influence on the positions taken by the TUC (Stevis 2018).[1] This chapter's analysis considers two dimensions: the role of labour activists within their own unions, within coalitions, and up through to union leadership and how unions from different sectors affect the positions of the TUC and the Labour Party. In fact, union activists have pushed the Labour Party to explicitly oppose fracking, a position taken under party leader Jeremy Corbyn in its 2017 Manifesto, even though the main energy union, GMB, is one of its biggest donors.

How can unions with a smaller occupational stake in fossil fuel employment negotiate for a withdrawal from fossil fuel with blue-collar workers and their labour organisations, which have an immediate interest in these potential jobs? Prinz and Pegels (2018) apply Kuzemko *et al.*'s (2016) power resources theory to explore which unions in Germany, for example, are 'forces for continuity' of fossil-fuel regimes and 'forces for sustainable change'. The authors incisively analyse the relative power of German union IGBCE, the main coal mining and chemical union, and ver.di, a service union and a rival for representing the public sector workers within these same industries. Prinz and Pegels analyse the 2017 Congress of the DGB, the German trade union confederation representing six million union members, when IGBCE and ver.di were able to prevent the Congress from making a decision to exit coal. The authors argue that when unions are more homogeneous in membership, as IGBCE is, they are able to act more forcefully and have more power. Ver.di represents a wider range of workers and their power is diffused; for that reason as well as others, union leadership sided with IGBCE. But Prinz and Pegels note that as the fossil fuel industry inevitably shrinks, IGBCE either shrinks as well or expands its diversity into different sectors such as in renewable energy, and the organisation may change its climate positions.

That more homogeneous unions exert greater influence is an intriguing argument. In 2018 there was another test case when the DGB Congress voted to uphold more stringent German climate goals with emission goals allocated to each sector (Kreuzfeldt 2018). According to Kreuzfeldt's 2018 reporting in the journal *Taz*, union leadership opposed higher emission goals but because of successful internal petition mobilisation, ver.di voted to uphold the higher goals. The German context is complicated (especially because of workers' co-determination on corporate boards), but the example is useful for highlighting how unions in different sectors and members of these unions can play a powerful role in shaping policy and for understanding how the shifting economy can affect the clout of fossil fuel unions within the labour movement.

In the UK, the Public and Commercial and Service Union (PCS), the Bakers Food and Allied Workers Union (BFAWU), the Fire Brigade Union (FBU), and several others are explicitly anti-fracking: they display their positions on their websites and work to put them into practice. Unite the Union and Unison, two other British trade unions, are more closely associated with energy workers even though they are heterogeneous; they both hold anti-fracking positions, but experience a more contentious struggle around the fracking issue. Because the General Municipal Boilermakers (GMB) and Prospect and Community unions have a sizeable number of members concentrated in the energy sector, they take a pro-fracking position. But labour activists from PCS and the Bakers are leading the charge against fracking along with members of the Unite and Unison and are pushing their unions to maintain and follow through on the anti-fracking position they have publicly taken.

While the TUC and the Labour Party respond to the occupational dilemma by pushing for green jobs, others call for a demand for climate jobs. The One Million Climate Jobs report (Neale 2014) was produced by the Campaign Against Climate Change and is backed by eight national unions. The report calls for the government to develop, support, and employ workers in jobs that are not just green but that actively combat climate change as a form of public service. Trade Unions for Energy Democracy (TUED), an independent international group, promotes the concept of public/social ownership of energy as a way to build infrastructure and equity in the energy sector (Figure 10.1). Hampton (2015) calls this the 'class perspective' on climate change.

Who controls energy and how society is structured in the future is also a question in the discussion around a 'just transition', that is, a plan for addressing the shift to a low carbon economy. There is a vast literature on this topic

Figure 10.1 Two union anti-fracking logos. Left: UK's PCS; right, the international TUED.

Source: reprinted with permission of PCS and TUED.

(Rosemberg 2010; Newell and Mulvaney 2013; Heffron and McCauley 2018; Smith and Patterson 2019), and the question of what employment will look like in a fossil-free economy lurks behind the struggle against fracking in the UK and indeed in all energy conflicts.

Fossil fuel industries in many instances create a plethora of jobs: they pump up local economies and permeate local cultures. The economic and social fabric of a society is dependent on workers at the point of production in a fossil fuel society for fuel and for energy. Mitchell (2011) argues that mining was critical to the rise and power of the labour movement, together with the industries that it made possible. In particular, coal required many people to physically cut the carbon from the depths of the earth and move it on to its destinations, people whose labour was required to make the economy work. For Mitchell, miners, rail, and port workers, known in the UK as the Triple Alliance, were the initiators of pivotal strikes and became the workers whom capitalists feared and maintained in their cross-hairs.

The coal union in the UK, the National Union of Mineworkers (NUM), is a prime example of a powerful political and cultural organisation, so powerful that in the 1980s Margaret Thatcher's government waged war against it in order to change course on subsidising coal. In a country where 200,000 workers were employed in the mines in the early twentieth century, that number declined to 170,000 in the 1980s and within a decade, to just a few thousand (Turnheim and Geels 2012, 36). The vivid memories of the bitter coal strike in 1984–1985 linger in the psyche of many UK communities that remain impoverished today as well as in the labour community in general. This strike was not just a defence of an outdated technology. It was a fight for the right to organise and a push for a socialist government that represented the interests of the workers. Support for the miners galvanised people to form solidarity groups throughout the country to take food to miners' communities, join their picket lines, and descend on urban centres in huge numbers to elevate the miners' demands of the Thatcher regime. Her administration dubbed the miners 'the enemy within' and launched a protracted attack on the union that included surveillance, secret police, physical attacks, and a widely propagated narrative against the labour union itself (Milne 2014).

The UK's transition away from coal was accomplished without concern to the well-being of mining communities. It devastated swathes of the UK that still have the highest rate of poverty in the country. The shutting of coal mines symbolises the concerns that workers in the fossil fuel industry have today about moving to a low-carbon economy.

Fracking has the potential to extend the status quo of the fossil fuel economy. The coalitions of residents, environmentalists, and trade unionists who are organising against fracking argue that climate change requires a drastic change rather than a continuation of a destructive greenhouse gas economy. Many of them are not just against fracking: they are arguing for climate justice and for a new society that is equitable for everyone on the planet.

Until this point, I have set the stage to tell a story about labour and climate change in the UK. I now turn to a discussion of the research methods that I used during the six months I worked in and around Liverpool to gather material for a study of labour activism against climate change and fracking and British labour unions. In this second section, I document the ideas of labour officers and activists I spoke to and events I attended. This account is a glimpse of the intense activity around climate change that labour unions in the UK are engaging in today.

The third section pieces together a narrative using interviews, Facebook event pages, newspaper articles, and scholarship to trace how union activists concerned about fracking began to organise with communities and environmentalists. This section also documents how the activism in climate camps moved into union spaces and influenced union resolutions. It also explores some of the differences among the unions and how occupational concentrations of energy workers within the unions affect their positions.

The final section briefly examines various union positions on fracking and the dialogue they are having about energy jobs and a just transition. What concrete ways are labour activists using to broaden the debate and using conferences, symposia, camps, and direct action to halt fracking? How are they using the resolution process to push towards banning fracking and what is the resistance to that effort?

Gathering materials and meeting people

I spent most of spring 2018 in Liverpool. Representatives of four different unions – PCS, Unite, BFAWU and GMB – spoke with me at length, participated in classes on labour and climate change that I taught at the University of Liverpool, and agreed to let me record their words.

I came to know the UK's union culture and the labour movement by various means. My own union at the University of Liverpool was on strike during the semester I was there, and speakers from many unions spoke at our rally in the famous CASA pub, an establishment founded by striking dockers in the 1990s. I learned first-hand about the restrictive laws the Tory government had passed to make picketing more difficult. I visited Unite the Union's halls in Liverpool and Salford and talked with and gave presentations on climate change and labour to Unite stewards and health and safety representatives. I also visited the GMB regional office in Liverpool and spent several hours talking with the local leadership about fracking. In addition, I met and had a long talk with a Unison anti-fracking activist. Activists in PCS and the BFAWU organised events or participated in events that I attended.

Clara Paillard lives in Liverpool and became my guide for much of what I experienced. I went to London for the Jobs and the Climate conference conducted by the Campaign against Climate Change. I learned about the One Million Climate Jobs campaign from Jonathan Neale, who wrote the report discussed above and met with Sarah Woolley of BWAFU and Chris Baugh, Assistant General Secretary of PCS. I also met Sam Mason, the PCS policy officer, and Barry Gardiner, Labour Party Shadow Secretary for Trade and Climate

Change, and also spoke at the conference. I also attended an environmental representatives' workshop to hear members from several unions speak about their activities.[2]

I visited the GMB headquarters in London to attend a symposium on air pollution as a Trade Union issue. There I met Colin Potter, a researcher from Unite, and Graham Peterson of the University and College Union and Secretary of the Greener Jobs Alliance[3] (a group bringing unions, students, and campaigners together). I also spoke at length with Hilda Palmer from the Manchester Hazards Campaign, one of a network of resource centres that performs educational work around health and safety, who works closely with labour.

I also went to People, Pits and Politics, a day of workshops that took place the day before the Durham Miners Gala in July 2018, an annual celebration of the miners' culture. The most striking workshop, which discussed climate change, was organised by Clara Paillard, in which she and Isabel Tarr, an anti-strip-mining activist, and Sarah Woolley, the regional officer for BFAWU, held a dialogue with two former miners, NUM activist David Douglass and a representative from the Orgreave Truth and Justice Campaign. John McDonnell, the Shadow Chancellor of the Exchequer, gave a speech to a packed room that included a socialist perspective on climate change.

I was invited to a two-day workshop organised by TUED: 'Reclaiming the UK Power Sector to Public Ownership: Developing a Programme of Action'. TUED plays a critical role in the UK and internationally in stimulating progressive thinking, policy, and action among trade unions. The workshop was held in June 2018 at Wortley Hall outside Sheffield, a former mine-owner's mansion that unions had rescued from collapse. Each room of Wortley Hall is named for a trade union, and labour images cover the walls to remind guests of workers' power. The workshop was a dialogue among participants who held different views about the energy industry but similar ideas on the need for public ownership. Representatives from many unions in the UK and the English and Scottish TUC spoke on the issues facing both the UK Labour Party and Europe. Presenters included Philip Pearson, retired TUC, Mika Minio-Paluello, energy researcher for the Labour Energy Forum, and Sam Mason, PCS. Roland Fulke from transform! europe spoke on the contemporary topic of municipal control of energy. Labour representatives and climate change activists circulated and discussed their latest briefings and reports. Not many rank and filers were there. PCS distributed their 'Energy Policy and Just Transition' report, Unison sent out its 'All or Something', the call to take into public ownership the customer and retail operations of the 'Big Six' energy suppliers (British Gas, EDF Energy, E.ON, NPower, Scottish Energy and SSE). GMB shared their Energy Briefing, 'March 2018', a document that details their support for fracking in the UK. There were other speakers on European climate issues, including Asbjorn Wahl, advisor to the Norwegian Union of Municipal and General Employees, speakers from the Transnational Institute, and other participants from Greece, Norway, Germany, and the Netherlands. Rebecca Long Bailey, Labour Party's Shadow Minister for Business, Energy and Industrial Strategy spoke to the participants about the Labour Party's energy platform. Susi Scherbarth from Friends of the Earth, Europe, offered the environmentalist perspective. During a break in the proceedings, Clara Paillard urged those of us who opposed fracking to come outside for a photo. She brought her neon green, handmade arty 'unions against fracking' poster for us to crowd around.

Aside from the participatory observation and interviews and workshops, I studied the websites of the various organisations, Facebook pages, Twitter feeds, and blogs of the unions and many individuals I met or to whom I was virtually introduced. I read PCS's publications 'Greening the Workplace' and 'Energy Democracy and a Just Transition', TUC's 'Greening Up', TUED's working papers series, and Unite's anti-fracking toolkit, which is hosted on the Sheffield Climate Alliance webpage. I also attended a number of labour and climate change symposia in the UK and on the continent, meeting scholars with deep roots in this field.

Apart from these many events, interviews, and talks, I learned how much deep conversation takes place in the UK's legendary pubs after the main event. This is where people often work out their issues or exchange ideas and feelings and accept each other as colleagues and potential allies within the labour and social justice movement.

What role do labour activists play in bringing unions together with communities and environmentalists?

This section describes how workers became bridge builders between residents who fear and despise the effects of fracking, environmentalists who oppose it, and union members who have become educated about the negative impact of the process and involved in the battle against it. Structurally, UK unions have developed a community membership that has given an opening and legitimacy to coalition-building (Hampton 2015).

I met activists involved in coalition-building who were members of the PCS and the BFAWU; Unite the Union also has specific community union officers in the Northwest. Reclaim the Power was the first environmental group I came across working with the Unite community anti-fracking events. I learned about this work from union webpages, through social media, personal communication, visits to the union halls, and then through published documents. I did not attend any 'frack-free' events but there were many such protests during my stay in the UK.

Despite the gentrification of many of its cities and the eroding of the country's industrial base, working class consciousness is still part of the British culture. While British labour has shrunk from almost 50 to 25.6 per cent (Bradley 2014) of the working class in Britain since 1979, this is still a high union-density percentage for a Western country. The context of British unionism, with its deep community-engaged unionism, is similar to the 'social justice unionism' discussed in US labour circles (Fletcher and Gapasin 2008) as a movement that builds reciprocal relationships between the community and union. The integration of climate change into the agenda of trade unions in other words relates to a shift toward community unionism (Lipsig-Mumme 2003; Wills 2001), or social movement unionism, and the organisation of worker alliances beyond specific sectors and workplaces (Lundström 2017, 6; Hrynyshyn and Ross 2011; Snell and Fairbrother 2011).

The British union culture has a self-reflective structure that promotes anti-racist caucuses as well as LGBT and disability caucuses; I cannot speak deeply to these as I did not attend any of their meetings. I did observe that Unite and GMB and other unions were invested in the 'Abortion is a Trade Union' issue in Northern Ireland and campaigned to strike down 'Paragraph 8', the Eighth Amendment of the Constitution of Ireland that criminalised women's right to choose. Marking social causes, such as abortion and climate change, union issues is more acceptable in the UK than it is in, for instance, in the US, but this does not mean that racism and gender discrimination, ableism, and climate denial are not present in the UK trade union movement.

Paillard was involved in pushing fracking to the centre of union debate and continues to be engaged in many efforts to bring labour into the organising efforts against fracking.[4] Her narrative is the main thread of this section. Lundström (2017) identifies labour leaders who are driven to transform their unions to be more climate-conscious with Gramsci's notion of the 'organic intellectual'. Gramsci (1995) describes those who through their experience with struggle gain knowledge of what needs to be done to mobilise members for change. Paillard is such a person. She plays a key role incorporating strategies for mobilising collective action on the rank and file level and uses the democratic forums within unions to push for policies to address climate change. Paillard describes herself as a lay leader in PCS: she emphasises how important it is that ordinary workers, not only officers, are involved as change agents. She uses strategies that are models to learn from to involve and collaborate with rank and file union members and community residents, such as showing up, building relationships, and encouraging respect, action, and creativity.

Paillard moved to Liverpool from her native France sometime in the early 2000s. As early as 2004 she was trying to get a Globalise Resistance group organised in Liverpool. She met Frank Kennedy from Friends of the Earth and suggested that they use the restaurant where she was working for their Climate Crisis meeting. A speaker from Globalise Resistance, Phil Thornhill, who is also from Campaign against Climate Change, and Frank Kennedy, who was the North West Coordinator for Friends of the Earth, were on the programme. Paillard has been an environmental/green rep in PCS since 2009 and Branch Chair for National Museums Liverpool. As a green rep, she was involved in working on *One Million Climate Jobs* with Jonathan Neale. She served on the executive committee of the Northwest Region Trade Council and then as Vice President of the PCS Culture Group. Paillard learned about fracking in France, heard about the earthquake that took place in Blackpool in the UK in 2011, and became determined to get unions involved in the anti-fracking struggle.

Dangers from fracking gradually became evident after licences to explore for shale gas were granted for many sites in the UK in 2007. The exploration and production company Cuadrilla received such a licence to explore for gas in Preston New Road in Lancashire. In 2011, soon after Cuadrilla began drilling, there were two tremors at the site, which were tied to the fracking exploration. The earthquakes registered 2.3 on the Richter's scale in April 2011 and 1.5 in May. Residents were awakened by the noise and the shaking, became worried about subsidence and their own health and safety, and began to inform themselves more fully about the potential negative consequences of fracking (Bradshaw and Waite 2017; Short and Szolucha 2017). 'Two earthquake tremors in north-west England earlier this year [2011] were probably caused by controversial operations to extract gas nearby, a report by the company responsible concluded' (Jowit and Gersmann 2011).

In the aftermath of the earthquakes, Paillard learned that anti-fracking campaigners were setting up Camp Frack I in a farmer's field near Southport close to where Cuadrilla had been granted additional licences to drill. The camp took place on 16–18 September 2011 (Siddle 2011), and was the sixth climate camp in the UK.

Climate camps are a means for environmentalists to bring media attention to a climate issue. They are also a way to educate people who come together with different kinds of knowledge about climate change. There are antecedents of climate activists and union activists working together, although these groups were bitterly opposed at first by trade union activists. The first UK climate camp was set up at Drax, in Yorkshire, in 2006, at the foot of a major coal-fired power plant. The participants had their backgrounds in 'grassroots, anti-capitalist activism combining social justice and anarchist perspectives' (Schlembach 2011, 1). In 2008, protesters set up a camp at Kingsnorth on the Hoo Peninsula in Kent against the possibility that a plant there would be fitted with Carbon Capture and Storage (CCS), a plan they considered 'a techno-fix that was unachievable and undesirable' (Schlembach 2011, 5). The participants claimed that coal itself was not the object of their protest, rather, it was capitalist growth. But they were met by a critique of their anti-working class politics from David Douglass, an anarchist and coal miner who had been a strike leader in the 1984–1985 conflict who advocates for clean coal using CCS in order to save jobs.

Climate activists went on to meet with other trade unionists at a 2008 conference sponsored by the National Union of Mineworkers and the Rail, Maritime and Transport Workers, and the Industrial Workers of the World. They were introduced to the idea of a worker-led 'just transition'. Some also became part of Workers Climate Action, a group that participated in a well-publicised and important worker-led occupation of the Vesta's wind turbine factory on the Isle of Wight in 2009, when the company laid off employees in preparation for closing the plant (Hampton 2015).

The Campaign Against Climate Change helped to organise Camp Frack 1 with help from a local group, Ribble Estuary against Fracking, a national group,

Frack Off, and Climate Camp groups from Liverpool and Yorkshire, Friends of the Earth, and the Cooperative Society. The next few months saw direct action as a number of campaigners tried to blockade and otherwise halt work on various other sites that Cuadrilla was exploring. The local Lancashire Council eventually had to make a decision about drilling in its territory, but that didn't happen until 2015, when the Council voted to reject Cuadrilla's application to plan fracking operations.

Paillard got in touch with Tina Louise Rothery, a small business owner who became involved in protesting against fracking in Preston New Road. From 10–12 May 2012, they both participated in Camp Frack 2. The Facebook event page advertised the event as a coalition of anti-fracking, Trade Union, and environmental groups. Merseyside and Cheshire Association of Trade Union Councils, the group that had elected Paillard their green representative, was also listed as co-sponsor. Stephen Hall, a labour activist and elected officer of Unite the Union, the Manchester Trade Unions Council, and Wigan Socialists, also attended Camp Frack 2. Judging from the photographs and news stories of the event, it was a wet camp, but enthusiastic.

Meanwhile, Paillard worked with the Green Party in Blackpool to organise an evening event at the Friends' Meeting House on 15 May 2012, entitled 'Fracking and One Million Climate Jobs'. Speakers were Philip Mitchell from the Blackpool Green Party and Anti-Fracking campaign and Paillard herself.

> Many who came were not in the trade unions. There were a few former dockers who were community-based and anti-racist and anti-austerity. There were allies in Manchester. Some individual trade unionists who were climate conscious and were trying to argue that fracking was a trade union issue.
>
> (Paillard, personal communication, 28 September 2018)

Ian Hodson of BFAWU was also at the event; Paillard told me that she had invited him to a workshop on food where she gave a presentation on the impact of fracking on the food chain that was eye-opening for Hodson. As a result, the Bakers union became very involved in fighting fracking, especially in the North-west, and also came on board with One Million Climate Jobs.

Together with her colleagues, Paillard saw the opportunity to push the issue in the Trades Union Congress.

> I was on the Northwest TUC executive and I was leading on climate issues. They were keen on having an event on climate change because they were into greening the workplace. No union was talking about just transition yet. Just the danger of the new [fracking] methods.
>
> (Paillard, personal communication, 29 September 2018)

The TUC found an article stating that the company IGas had been given a licence to drill between Manchester and the river Mersey/Dee estuaries, and that abundant shale gas had been found in the Wirral, a rural part of Liverpool. In 2012, the Labour and Green party councillors on the Wirral Council voted to ban fracking (Barnes 2012). Paillard notes that at this time progress on the fracking issue was happening but that it was uneven: 'fracking was starting to be on the political agenda but not on the union agenda'. When someone in the Northwest TUC decided they needed an anti-fracking motion, Paillard wrote it, and at the July meeting the motion for a frack-free zone passed. 'I argued that it should be put forward to the national body. Each council could put two motions forward, and it was supported', Paillard said (Paillard, personal communication, 29 September 2018). This motion was the one that would go to the full TUC for a vote in September 2012.

On 25 June 2012, Paillard posted a notice for another event: 'A Northwest TUC conference will be held on 21st July in Blackpool with workshops on fracking and other environmental issues'.[5] The event was an important opportunity to build union consciousness and interest in the anti-fracking movement. Called 'For a Future that Works: Trade Unions & The Environment, A NW TUC Conference', the conference description integrated concerns about fracking and climate change into equity issues:

> Over the past 50 years our current global economic model has failed to deliver the sustainable society that we need. Poverty has increased in absolute terms. The gap between rich and poor is growing and at the same time, we have already breached the 'safe operating limit' of our planet. We need to build a new sustainable and socially just economy.
>
> (2012 NW TUC Conference announcement)

The 21 July conference was a large step forward for anti-fracking labour to define its reasons for opposing the drilling. Bron Szerszynski blogged[6] about the event, providing rich insight into the trade union climate change thinking and how deeply political it is.

This series of labour engagements in anti-fracking camps and conferences highlights the importance of labour activists organising within the movement and in coalition to have an impact on the perspectives of national organisations. In September 2012, the TUC passed a series of key energy resolutions, including the moratorium on fracking. This moratorium vote reflected a surge of union interest in the issue. 'We won the argument during the debate', Paillard said. The TUC passed the moratorium based on 'the precautionary principle that many of the risks, such as fugitive methane emissions, had not been adequately quantified or controlled' (Hampton 2015, 113).

The anti-fracking movement grew throughout 2013. Paillard noted that organising in the Northwest was largely done by working-class folks. In Balcombe in the South, anti-fracking organising that flourished that year was based on the middle-class, she said, although there was some trade union involvement (Paillard, personal communication, 29 September 2018). There was direct action timed to coincide with the annual Tory Party conference; many were arrested and it did get a lot of press. In September 2013, a large march was organised by the Manchester TUC called 'Frack the Frackers, Climate Jobs Now'. In August 2014, Reclaim the Power, a group which advocates for solutions to climate, economic, and social crises, held a climate camp in Blackpool that brought many different organisations together, including trade unionists. The workshops included one on energy workers' struggles in the UK, from fossil fuels to renewables.[7]

How do energy unions and unions on the periphery of the fossil fuel industry share power in making decisions on fracking and related issues within the TUC and the Labour Party? Many of the unions who have a concentration of energy workers also have a good mix of members from other industries but see fracking as a potential area for recruitment and growth. 'Within the TUC the industrial unions were pushing energy mix [including fracking], and it was a potential area where GMB could get members' (Paillard, personal communication, 29 September 2018). In 2014, according to Hampton (2015) 'the Trades Union Congress agreed to continue to consult affiliates about a just transition to a low-carbon economy, including a moratorium on extreme energy such as shale gas extraction (fracking)'. The debate was contentious. Tam McFarlane (FBU), summing up the 2014 debate, pointed to the 160 anti-fracking organisations across the UK, arguing that unions should be 'on the side of people who are on the same political side as us' (Hampton, 2015, 113). Hampton goes on to document this momentous year.

> Both the Unite and PCS conferences in 2014 passed resolutions calling for a moratorium on fracking. However, there was considerable debate around the resolutions, with both motions yielding an explanation from the TUC general Council that the position did not preclude support for the gas industry as a whole.
>
> (Hampton 2015, 113)

In other words, the moratorium was becoming increasingly difficult to uphold among all the union affiliates.

A 2014 article in the *Socialist Worker* describes the vote at the TUC conference in more detail, stating that the Executive Committee of Unite had put forward a motion 'softening' the anti-fracking stand (Sewell 2014). TUC Executive Committee members argued that although the union was opposed to fracking, it was a reality and the union should go out and recruit the workers. Rank and file members, including a health worker whose son had been arrested at an anti-fracking protest, mocked the officer's motion. 'Fracking is morally abhorrent. We should be putting this money into renewables, not more fossil fuels' (Harris in Sewell 2014). Members voted down the Executive Committee's motion.

When I asked Paillard if there were any energy workers who were climate activists, she reflected that perhaps trade union climate advocates may not have done enough to engage energy workers. 'We were trying to get unions involved in general, not energy in particular'. She said that, in her opinion, the lack of the rank and file and climate activists who are sent to conferences to represent and vote on behalf of their unions distorts the concerns of working people in unions (Paillard, personal communication, 28 September 2018).

> In 2014 when the Unite conference took place, there were three motions with fracking. One came from the energy sector. My feeling was that the energy sector was always represented by the officers – not the workers. PCS when there were climate change events – they always had lay reps like me, while UNITE usually has officers attend. One of the problems – officials are dealing with the strategic work and they don't consult with their members. That's the epitome of how UNITE worked … anti-fracking from lay leaders, while national exec was pro fracking.
> (Paillard, personal communication, 28 September 2018)

The involvement of rank and file workers speaking with workers directly about climate concerns, involving them in education, and developing the leadership and voice of workers in decision-making spaces is critical to Paillard:

> I think there's some contempt – they [the officers] know better and we need to have a more mature conversation.… They are interested in a technological solution that may be short-sighted – like CCS or hydrogen. They don't explore with an open mind. The technological solutions have been put forward by industry. Because they negotiate with industry, they [the officers] develop relationships that are based on mutual trust. Officers are far too amiable with employers and listen to them over the climate movement. That's my view.
> (Paillard, personal communication, 29 September 2018)

This perspective resonates with the analysis Lundström (2017) uses when he contrasts the bottom-up practices of the PCS in organising climate change consciousness to what he characterises as top-down organising in Swedish unions.

The Lancashire County Council voted on fracking in 2015 in the midst of the controversy, and the planning application was rejected. Locals, environmentalists, and anti-fracking unionists were overjoyed but their joy was short-lived.

> The decisions were then appealed by Cuadrilla and there was a public enquiry in February and March of 2016. On 6 October 2016, the central Government overturned the initial decisions at one site and gave Cuadrilla more time to address traffic concerns at the other.
> (Bradshaw and Waite 2017, 29)

Several scholars suggested that this was an unethical process:

> For example, for much of the public fracking debate so far the government was deciding energy policy with the assistance and advice of the 'lead

non-executive director' at the Cabinet Office, Lord Browne, who just so happened to be the Chairman of shale gas company Cuadrilla Resources [the corporation applying for the fracking license]. If this situation arose in a developing country, it would likely be described as corruption.

(Short and Szolucha 2017, 11)

The development of fracking was turning into an overturning of local democracy by the central government at the bidding of corporate interests.

Protests picked up again after the 2016 decision, and they are still going on as of this writing. People have continue to camp out at Preston New Road. Fracking now requires a more streamlined process, and sites in several parts of Britain are being tested. Unionists continue to be involved. There are special union days at Preston New Road, for example, when individual unionists are encouraged to come with a group of their mates and bring their banners. One of those days was 9 June 2018 (Figure 10.2).[8] The Facebook event page announced:

We are Trade Unionist and workers against fracking. A growing number of Trade Unions across the UK and globally have been supporting resistance against fracking and the dash for gas. We support Climate Jobs, not Fracking Jobs! We stand in solidarity with antifracking campaigners in the community and will be going to Preston New Road in support of the people of the Fylde, the Nanas and all the other amazing people who are resisting every day at the site. Bring your Trade Unions banner, get your union, your branch, your Trades Council to send representatives and message of solidarity.

(Facebook event page, 2018)

Members of Unite succeeded in enlisting their union to produce a booklet on how to campaign against fracking. Published in June 2018, the Unite anti-fracking toolkit includes an introduction written by Unite's Assistant General Secretary Gail Cartmail. It includes a detailed discussion of the problems of fracking and answers to the myths that the industry perpetuates. There is a section on how important it is for Unite to build strong ties with communities, how fracking affects many people, and how to start an anti-fracking community organisation. The text is eloquent, stating that it may be hard to stop a single fracking well, but in the face of communities coming together against the industry, it is harder for companies to feel confident about future financial rewards. There are also well-thought out instructions for how to understand what happens during the planning process of a proposed fracking project, how to organise a protest, and how to write a press release. There are maps of licensed areas all over the UK as well as lists of useful articles and many other resources, including how to pass a motion in your local branch. But this toolkit is difficult to find: it does not appear to be searchable through the Unite webpage. It can be found on the Sheffield Climate Alliance website.[9] More information on the

Figure 10.2 Unions and community groups demonstrate at Preston New Road in Lancashire on 9 June 2018.

Source: photo by Neil Terry, reprinted with permission.

struggles of rank-and-file Unite members can also be found on what is called the
frack-free Cleveland blog, produced by a group that focuses on protests around
the Kirby–Misperton North Yorkshire fracking site.[10] Unite's website lists the
policies passed at the last four conferences, and the anti-fracking policy is clearly
included.[11]

BFAWU continues to grow its social justice unionism with the 'fight for 15 £
per hour', which extended McDonald's workers' strikes to Wetherspoon
employees in recent years. The union sees fracking as threatening both the
water table and members' livelihoods. A discussion of the motion put forward
by the Bakers at the 2018 TUC conference was published in their 'Green Stuff'
newsletter, as well as in the union's magazine, *Foodworker*. Ian Hodson summarised the TUC's approach to the fracking issue at the Conference:

> Another area where the BFAWU has enhanced its reputation is with the
> green agenda and once again Sarah Woolley gave a compelling speech to
> the congress, very much in line with our own Conference policy on frack-
> ing, although as previously stated it was watered down somewhat by the
> amendment from Unison. This did not stop the momentum and commit-
> ment to the environment as Sarah spoke at a fringe [workshop] that was
> well attended, where another opportunity for delegates to hear about the

BFAWU's progressive green policies was aired and supported. Fracking – motion 8, I was proud to move this motion in which we called for fracking to be banned and for the TUC and affiliated unions to work on energy policy, decarbonisation and develop a strategy to support workers transitioning to a greener, zero carbon economy. UNISON put in an amendment which in our opinion diluted the motions and potentially allowed for the TUC to support fracking in the future. We didn't accept the amendment, however congress voted to accept it and the motion was carried, albeit weaker than our original intention.

(*Foodworker* 2018, 4)

For the Bakers' union, getting out in front of fracking and moving the moratorium to a ban is a proud achievement.

Hodson also discussed a shift in how policies around energy may be made in the future by the TUC, which has since been echoed in a recent Labour Party decision to form a working group of energy unions:

There were a number of motions that were passed which caused concern to a number of the smaller unions including ours. One of these was motion 7 moved by the GMB which in a nutshell called for only GMB, Prospect, Unison and Unite to be able to develop policies on energy, industrial strategy and climate change as they are the only unions to have members working in the energy sector. We believe this is the wrong move; energy affects everyone, energy costs, environmental and climate change impacts every sector, industry, business and person not just the workers within the energy sector, therefore everyone should have the opportunity to develop policy. Unfortunately, with three of the biggest unions in the movement in support of the motion it was carried, however in my opinion this is dangerous and divisive. In our current environment with the government we are under, division is the last thing we need!

(*Foodworker* 2018, 5)

Sam Mason from PCS reiterates the importance of all unions, not just the energy unions, having a say in climate and industrial policy. She underlines the importance of protecting energy workers,

but if the party really want to lead a revolution in green jobs and industrial strategy, they need to open their eyes to a transformational whole economy approach that takes account of all workers in that transition. An economy that recognises the need for Labour values of justice, equality, solidarity, and democracy across all sectors in the Just Transition debate.

(Mason in Levy 2017)

PCS, like BWAFU, argue that all unions are affected by the transition away from fossil fuel and all should be involved in decision-making on Labour Party climate policies.

Conclusion

The labour movement plays a critical role in developing climate policy; as a key issue, the debates on fracking illustrate serious divisions on the issue within labour. The Trades Union Congress passed a resolution for a moratorium on fracking, but there is tension between the unions based directly or indirectly in the energy field. Activists who oppose fracking have been persistently organising with community and environmentalists because of the negative impacts of the fracking process and the delaying effects of the development of gas on the use of renewable energy. Since 2015, unions have had increasing differences over fracking, with energy unions supporting it and progressive unions such as PCS and BWAFU in the forefront of opposing it.

Fracking is highly controversial; it is strongly associated with environmental harm. The moratorium on fracking passed by TUC in 2012 as a result of the work of activists and anti-fracking unions is still in effect. The Labour Party's environmental policy paper, 'Green Transformation' (n.d.), maintains a strong anti-fracking position despite the fact that their biggest donor is the pro-fracking GMB. By virtue of its organising a working group of energy workers to formulate climate policy, the current position of the Labour Party raises the question of who should be making energy policy. Should everyone, that is, communities and workers directly and indirectly involved in fossil fuel industry, have a say in deciding energy policy? Will unions outside the energy circle gain a stronger influence because of their numbers and the strength of labour activism within their ranks? Will those who have doubts about fracking within energy unions affect their unions' vision?

Labour activists working in coalition with anti-fracking communities are raising their voices and pushing their leadership to take a stronger stand to promote climate jobs over fossil fuel employment. Explicit support from the leadership of unions such as PCS, BFAWU and the FBU contributes to developing leaders from the rank and file. How UK labour and the Labour Party develop a vision and a strategy for a just transition in the face of climate change is very much in the balance.

Notes

1 There is variability in this internationally, with some unions, even those concentrated in energy, affiliating with environmental positions, especially if there may be employment for members. The unions involved in the Blue Green Alliance are an example.
2 To read more about the conference, see Jobs and climate: Planning for a future which doesn't cost the earth: www.cacctu.org.uk/jobs_and_climate
3 See their webpage at www.greenerjobsalliance.co.uk/
4 Personal communications with Clara Paillard over a year, November 2017 to November 2018.
5 All quotes and comments on this conference come from: www.facebook.com/events/449179215105987/ (accessed 31 October 2018).
6 Bron's blog can be read at: https://csecblog.wordpress.com/2012/07/22/trade-unions-and-the-environment-conference-blackpool-21-july-2012/ (accessed 31 October 2018).

7 For the full programme, see http://reclaimthepower.org.uk/2014campblackpool/2014-campprogramme/

8 See Facebook event www.facebook.com/events/164196737583363 (accessed 31 October 2018).

9 www.sheffieldclimatealliance.net/2018/10/unite-the-union-solidarity-with-anti-frackers/ (accessed 31 October 2018).

10 http://frackfreecleveland.co.uk/index.php (accessed 31 October 2018).

11 https://unitetheunion.org/who-we-are/policies (accessed 31 October 2018).

References

Barnes, G. (2012) Wirral Safeguard pledge over controversial oil drilling. *The Globe*, 9 July. Available at: www.wirralglobe.co.uk/news/9806178.Wirral_safeguard_pledge_over_controversial_gas_extraction/?fbclid=IwAR13IlUEc8Or1pkQI8Eh8nafgiq3zEtA5z7wwomJpF8mfu29X6VvjQPIfLc

Bradley, M. (2014) Trade Union density in Britain. *Socialist Review*. Available at: http://socialistreview.org.uk/396/trade-union-density-britain (accessed 31 October 2018).

Bradshaw, M. and Waite, C. (2017) Learning from Lancashire: exploring the contours of the shale gas conflict in England. *Global Environmental Change*, 47, 28–36.

Clough, E. (2018) Environmental justice and fracking: a review. *Current Opinion in Environmental Science & Health*, 3, 14–18.

Christopherson, S. (2015) The false promise of fracking and local jobs. *The Conversation*, 27.

Fletcher, B. and Gapasin, F. (2008) *Solidarity Divided: The Crisis in Organized Labor and a New Path Toward Social Justice*. Oakland, CA: University of California Press.

Foodworker (2018) *Autumn 2018*. Available at: https://assets.nationbuilder.com/bakersunion/pages/439/attachments/original/1537773329/FW_AUTUMN_2018_web_version.pdf?1537773329 (accessed January 2019).

Foran, C. (2014) How many jobs does fracking really create. *National Journal*.

Graziano, M., Lecca, P., and Musso, M. (2017) Historic paths and future expectations: the macroeconomic impacts of the offshore wind technologies in the UK. *Energy Policy*, 108, 715–730.

Gramsci, A. (1995) *Further Selections from the Prison Notebooks*. Minneapolis, MN: University of Minnesota Press.

Hampton, P. (2015) *Workers and Trade Unions for Climate Solidarity: Tackling Climate Change in a Neoliberal World*. Abingdon, UK: Routledge.

Heffron, R.J. and McCauley, D. (2018) What is the 'Just Transition'? *Geoforum*, 88, 74–77.

Hrynyshyn, D. and Ross, S. (2011) Canadian autoworkers, the climate crisis, and the contradictions of social unionism. *Labor Studies Journal*, 36(1), 536.

Jaspal, R. and Nerlich, B. (2014) Fracking in the UK press: threat dynamics in an unfolding debate. *Public Understanding of Science*, 23(3), 348–363.

Jowit, J and Gersmann, H. (2011) Fracking 'probable' cause of Lancashire quakes. *Guardian*, 2 November. Available at: www.theguardian.com/environment/2011/nov/02/fracking-cause-lancashire-quakes

Kreuzfeldt, M. (2018) Tightened application accepted: DGB introduces climate. Available at: www.taz.de/Antrag-beim-Bundeskongress/!5503943/ (accessed 16 May 2018).

Kuzemko, C., Lockwood, M., Mitchell, C., and Hoggett, R. (2016) Governing for sustainable energy system change: Politics, contexts and contingency. *Energy Research & Social Science*, 12, 96–105.

Levy, G. (2017) Moving the unions past fossil fuels. *People and Nature*. Available at: https://peopleandnature.wordpress.com/2017/08/09/moving-the-trade-unions-past-fossil-fuels/ (accessed 20 August 2018).

Lipsig-Mumme, C. (2003) Forms of solidarity: unions, the community and job creation strategies. *Just Policy: A Journal of Australian Social Policy*, 30, 47.

Lundström, R. (2017) Going green–turning labor: a qualitative analysis of the approaches of union officials working with environmental issues in Sweden and the United Kingdom. *Labor Studies Journal*, 42(3), 180–199.

Lundström, R., Räthzel, N., and Uzzell, D. (2015) Disconnected spaces: introducing environmental perspectives into the trade union agenda top-down and bottom-up. *Environmental Sociology*, 1(3), 166–176.

Maniloff, P. and Mastromonaco, R. (2017) The local employment impacts of fracking: a national study. *Resource and Energy Economics*, 49, 62–85.

Mayer, B. (2009) Cross-movement coalition formation: bridging the labor- environment divide. *Sociological Inquiry*, 79, 219–239.

Metze, T. and Dodge, J. (2016) Dynamic discourse coalitions on hydro-fracking in Europe and the United States. *Environmental Communication*, 10(3), 365–379.

Milne, S. (2014) *The Enemy Within: The Secret War Against the Miners*. London; New York: Verso Books.

Mitchell, T. (2011) *Carbon democracy: Political Power in the Age of Oil*. London; New York: Verso Books.

Neale, J., 2014. One million climate jobs: solving the economic and environmental crises. www.campaigncc.org/sites/data/files/Docs/one_million_climate_job\s_2014.pdf (accessed 25 October 2018).

Newell, P. and Mulvaney, D. (2013) The political economy of the 'just transition'. *The Geographical Journal*, 179(2), 132–140.

Obach, B.K. (2004). *Labor and the Environmental Movement: The Quest for Common Ground*. Urban and Industrial Environments. Cambridge, MA: MIT Press.

Prinz, L. and Pegels, A. (2018) The role of labour power in sustainability transitions: insights from comparative political economy on Germany's electricity transition. *Energy Research & Social Science*, 210–219.

Räthzel, N. and Uzzell, D. (2011) Trade unions and climate change: the jobs versus environment dilemma. *Global Environmental Change*, 21(4), 1215–1223.

Rosemberg, A. (2010) Building a just transition: the linkages between climate change and employment. *International Journal of Labour Research*, 2(2), 125.

Rousu, M.C., Ramsaran, D., and Furlano, D. (2015) Guidelines for conducting economic impact studies on fracking. *International Advances in Economic Research*, 21(2), 213–225.

Schlembach, R. (2011) How do radical climate movements negotiate their environmental and their social agendas? A study of debates within the Camp for Climate Action (UK). *Critical Social Policy*, 31(2), 194–215.

Sewell, D (2014) Unite union policy conference takes on NHS cuts, Labour, fracking and racism. *Socialist Worker*, 8 July. Available at: https://socialistworker.co.uk/art/38530/Unite+union+policy+conference+takes+on+NHS+cuts%2C+Labour%2C+fracking+and+racism

Short, D. and Szolucha, A. (2017) Fracking Lancashire: the planning process, social harm and collective trauma. *Geoforum*.

Siddle, J. (2011) Environmentalists join Southport protest camp against fracking shale gas drilling. *Guardian*, 19 September. Available at: www.theguardian.com/environment/2011/jul/22/activists-camp-frack-shale-gas?fbclid=IwAR0jQg-rAJX86MbqB5bHIo4plrPQcXXgS7DUofKBZhoyOgzZUfb5RnVaJmw

Smith, J. and Patterson, J. (2019) Global climate justice activism: 'the new protagonists' and their projects for a just transition. *Ecologically Unequal Exchange*. Cham: Palgrave Macmillan, 245–272.

Snell, D. and Fairbrother, P. (2011) Toward a theory of union environmental politics: unions and climate action in Australia. *Labor Studies Journal*, 36(1), 83–103.

Stevis, D. (2018) US labour unions and green transitions: depth, breadth and worker agency. *Globalizations*, 15(4), 454–469.

The Green Transformation: Labour's Environmental Policy (n.d.) Printed and promoted by Jennie Formby, General Secretary, the Labour Party, on behalf of the Labour Party.

Turnheim, B. and Geels, F.W. (2012) Regime destabilisation as the flipside of energy transitions: Lessons from the history of the British coal industry (1913–1997). *Energy Policy*, 50, 35–49.

Wills, J. (2001) Community unionism and trade union renewal in the UK: moving beyond the fragments at last? *Transactions of the Institute of British Geographers*, 26(4), 465–483.

Part III

Creating sustainable cities and infrastructures

11 Why does everyone think cities can save the planet?

*Hillary Angelo and David Wachsmuth**

Introduction: From the city as a sustainability problem to the city as a sustainability solution

While fears of global warming and environmental catastrophe loom ever greater, urban areas continue to expand unevenly. And, in the face of environmental and urban crises, the ideal of the 'sustainable city' increasingly takes a leading role in urban planning and policy discourse. Policy-makers focus on the vulnerability of cities – as global population and economic centres – to sea level rise, droughts, storms, and other impacts of climate change and on the role of the built environment in determining sustainability outcomes, particularly through influencing greenhouse gas emissions. Planners and pundits increasingly recognise the interdependence of local sustainability problems faced by cities around the world, and thus the necessity for a 'common front' to address these problems, alongside a widespread belief in the failure of national politics to address climate change. Social movements, activists, and scholars all target cities as locations for progressive visions of future sustainable life. Accordingly, sustainable urbanism has become a new policy common sense. As Susan Parnell (2016, 529) argues, 'there is no longer a question of whether cities are important for sustainable development, but rather why and how the urban condition affects our common future'.

In short, everyone now thinks cities can save the planet. But given the current ubiquity of political, cultural, and academic discourse connecting cities and sustainability, it is clarifying to observe that fewer than two decades ago the influential environmentalist Herbert Girardet (1999) was still posing the relationship between the two as a potential 'contradiction in terms'. What happened? Why does everyone think cities can save the planet, and why now?

The emergence of this policy discourse can be traced to the 1990s, when sustainable city initiatives began to emerge in different parts of the world (Wheeler 2000), following the broader emergence of sustainability thinking after the 1987 publication of the UN's Brundtland report and the 1991 Earth Summit (Brundtland 1987; Du Pisani 2006). Girardet (1996) was one of the most important pioneers of a specifically urban understanding of sustainable development, while Mike Jenks and colleagues made a contemporaneous argument for the 'compact city', motivated by the insight that variations in urban form are systematically linked to socio-environmental outcomes (Breheny 1992; Burgess and Jenks 2002; Jenks *et al.* 1996). The subsequent 'smart growth' movement applied the same insight to more suburbanised contexts in the United States (Burchell *et al.* 2000; Duany *et al.* 2010). And although the sustainability of urban density was still a relatively counterintuitive idea in 2004 when David Owen (2004) re-elaborated it for a popular audience, it has since become the foundation for municipal and national climate action plans around the world.

Current sustainability discourse reflects not just the assumption that cities can and should be green, but also that they are the most likely solutions to our global environmental problems. So even as suburban modes of urban growth have continued to expand and intensify across the globe in the last 20 years (Keil 2013; Hamel and Keil 2015), this period of time has also seen the emergence of something close to a policy consensus around the idea that urban density has environmental value, particularly with respect to the necessity of reducing greenhouse gas (GHG) emissions and associated energy use. Conceptions of the smart city involve many of the same assumptions – that the built environment can be re-engineered in ways that can improve both local and global sustainability trajectories – although from a new perspective of techno-optimism (Hollands 2008). The arrival of 'resiliency' as a new urban policy concept and buzzword reframed these sustainability imperatives again in the context of environmental crisis (Ahern 2011; MacKinnon and Derickson 2013). Today, in short, urban sustainability policies promoting green amenities and the environmental value of density have become ubiquitous and a vision of sustainability as dense, green urbanism is the norm.

We have previously argued that urban environmental research has been too narrowly focused within city boundaries (Angelo and Wachsmuth 2015), and that expanding the frontiers of urban sustainability offers a more spatially and socially inclusive vision of urban futures (Wachsmuth *et al.* 2016). Here, we add a temporal dimension to this intellectual project (see also Wachsmuth and Angelo 2018; Angelo and Vormann 2018) in order to interrogate a shift that otherwise risks being naturalised: from the city as a sustainability problem to the city as a sustainability solution. In this chapter, we provide a historical account of the emergence of a 'cities can save the planet' common sense, focusing on the trajectory of international urban sustainability policy efforts from the 1960s onward, first to address the environmental deficits of urban areas and eventually to harness urban political energy to improve global sustainability trajectories. This account identifies three major historical developments which collectively established urbanism as a plausible policy solution to global environmental concerns: sprawl, informal settlements, and climate change. Then, on the basis of this account, we lay out an agenda for contemporary urban environmental research which is historical, multi-spatial, political, and representational.

Sprawl, informal settlement, and climate change: the emergence of a global urban sustainability imaginary

What established cities as common-sense solutions for global environmental problems? We argue that three major historical developments that have their roots in the 1970s and intensified throughout the 1990s contributed to this discourse and policy shift. The first was the growth of urban sprawl in the Global North, especially the United States, whose endless expansion Northern environmentalists worried would threaten the natural environment and lead to unsustainable forms of resource use. The second was the growth of informal settlements in the Global South, deemed environmental problems by international development experts because of their rapid growth, physical precarity, and rampant public health problems. The third threat was climate change, which became a global problem connecting concerns across North and South as understanding of cities as key sources of greenhouse gas emissions – and therefore key arenas for reducing emissions – grew.

In the past few decades, there has been a parallel shift in proposed solutions to each of these socio-ecological problems. Urban and environmental thinkers and policy-makers have proposed first non-city solutions, and then city solutions, to all three. Following several decades of worries about sprawl's potential for endless expansion, a city solution to sprawl as an environmental problem emerged in the form of the compact city. By 2010, informal settlements even began to be proposed as a solution to environmental problems, as the former came to be seen as a kind of vernacular form of sustainability (flexible, impermanent, dense, and low-cost), and as models for 'resilient' urbanism. In addition – primed by urban environmental thinking regarding sprawl and informal settlements in the context of increasing awareness about the vulnerability of coastal cities and the failure of international treaty efforts – cities have been promoted to the role of premier anti-climate-change environmental protagonists.

Each of these problems highlighted the connection between cities and the natural environment, first in an oppositional relationship and then in a potentially harmonious one. Taken together, urban-environmental agendas that emerged in response to these concerns have added up to a decisive shift in the view of the city, from a sustainability problem (requiring non-city solutions) to the city itself as a solution to global sustainability problems. While, in the mid-twentieth century, cities were almost universally understood to be destroying nature, now, in the twenty-first, we see a widespread understanding of cities in harmony with nature and a vision of sustainability as specifically *urban* has become the norm.

Sprawl

Although it is associated, above all, with post-war United States, urban sprawl has been a leading imaginary of the city as a sustainability problem across a range of spatial and historical contexts. It is now broadly understood as an urban challenge worldwide, with affluent – often secessionist – suburbs and poor informal settlements both accounting for a disproportionate share of new urban development in Global South metropolises (UN-Habitat 2012, 11). Sprawl also has a far longer lineage than the standard US-post-war-centred history acknowledges: as Robert Bruegmann (2006) argues, low-density urban development has been a constant of urbanisation across space and throughout history. At the same time, the onset over the course of the twentieth century of automobility as a way of life (Sheller and Urry 2000) provided for a *generalisation* of urban sprawl in the parts of the world where car ownership became ubiquitous. Nowhere was this truer than the United States after the Second World War – the most car-dependent country on the planet, where suburban sprawl achieved its most extreme form. In the US, far more so than other advanced capitalist countries, suburbanisation played a central role in the broader economic regulation of post-war national capitalism (Walker 1981; Florida and Feldman 1988), and the US continues to be a global outlier with respect to the dispersed spatial form of its urban growth patterns (Schneider and Woodcock 2008). The US is thus a key site to document the historically changing relationship between urban sprawl and sustainability thought.

From the early days of industrial urbanisation into the post-war era, suburban development in the US was a commonly proposed *corrective* to the environmental deficits of urban life – from questions of poor health and overcrowding in polluted, densely packed tenements to racially uneven exposure to industrial pollutants (Pulido 2000). But the US, in addition to being the centre of post-war suburbanisation itself, hosted the first major popular and intellectual opposition to suburbanisation on environmental grounds, beginning in the 1960s (Rome 2001; Sellers 2012). Inspired by works such as Rachel Carson's (2002 [1962]) *Silent Spring*, Barry Commoner's (1971) *The Closing Circle*, and the Club of Rome's *The Limits to Growth* (Meadows *et al.* 1972), the growing environmental movement saw in suburban sprawl unsustainable patterns of resource consumption, land use, and pollution. Importantly, these critiques all tended to see suburbanisation as synonymous with urban development more broadly and therefore derived generic critiques of urbanisation from more specific characteristics of suburban sprawl. Moreover, the thrust of these critiques was, to varying extents, ultimately anti-capitalist as opposed to simply just anti-suburban. Ernest Callenbach's (2009 [1975]) influential novel *Ecotopia* is exemplary in this regard: it posits a near-future where the Northwest United States secedes from the rest of the country and establishes an ecologically sustainable polity. Cars are abolished, energy production is decentralised, and cities are partially returned partway to nature. But economic progress is also abolished and replaced with a 'stable-state' system in which human and non-human ecological imperatives are balanced. For such environmental critics of post-war urban development, the latter was understood as one expression (albeit an important one) of a broader socioeconomic system that was fundamentally incompatible with ecological principles. The 'solution' to the problems posed by this system was to dismantle both the economic engine (capitalist growth) and its spatial expression (sprawling urban development).

In contrast to previous waves of opposition to sprawl throughout the twentieth century, however, current anti-sprawl ideals of urban density and the compact city have become the concrete imaginary of the *city* as a sustainability solution. This shift began with the oil crisis of the 1970s (Girardet 2007), which foreshadowed the end of cheap oil and grew throughout the 1980s and 1990s as city governments began trying to attract the white middle class back from the suburbs by reinvesting in hollowed-out downtowns, and as concerns about pollution and carbon emissions grew.

Already by the mid-1980s, the architect Peter Calthorpe (1985, 1) argued that 'Ideally, the city is the most environmentally benign form of human settlement. Each city-dweller consumes less land, less energy, less water, and produces less pollution than his counterpart in settlements of lower densities'. The influential Australian urbanists Peter Newman and Jeffrey Kenworthy (1989) made a similar point in *Cities and Automobile Dependence*, while Jenks *et al.* (1996) brought the idea of the 'compact city' to academic and policy prominence several years later. By 1998, the Sierra Club had launched its 'Challenge to Sprawl' programme, with an accompanying report which explicitly cast suburban sprawl as an ongoing environmental catastrophe (Sierra Club 1998). What these perspectives have in common, which they share with more recent density-focused visions of green urbanism such as Owen (2004) and Glaeser (2011), is that they clearly distinguish suburban sprawl from dense urban development and locate environmental problems with the former and solutions with the latter. Even though small-scale communities scattered across a lightly urbanised landscape may still appear 'greener' than the dense 'grey' cities they surround (Wachsmuth and Angelo 2018), urban policy common sense now increasingly sees dense urbanism as the more sustainable choice.

Informal settlements

As far back as the beginning of the twentieth century, futurists were predicting the urbanisation of the world. But few would have predicted that, when global society began to concentrate decisively in cities in the second half of the century, it would do so in informal settlements, or 'slums', as they are often referred to by local organisations and the international development community (Satterthwaite 2016, 3). Early twentieth century urban slums were understood as by-products of industrial development and were expected to give way to formal settlements as that development progressed. However, by the 1970s, informal settlements were expanding rapidly in cities throughout the Global South and were largely unaccompanied by the predicted industrialisation. Owing to their apparently inexorable growth, these settlements also rapidly began to be seen as environmental problems – a perception that remained through to the end of the twentieth century. Mike Davis (2006, 134–142), for example, described two aspects of informal settlements' unsustainability in his apocalyptic *Planet of Slums*. They contributed to environmental degradation by polluting surrounding recreational space, agricultural land, water sources, and natural habitat with human waste and pathogen, and endangered residents lacking health care, infrastructure, and public services through exposure to 'natural' disasters such as flooding and landslides as well as contaminated food and water, and infectious diseases.

The framing of urban informal settlements as environmental problems is particularly visible throughout global policy discourse in the final quarter of the twentieth century. The 1976 Vancouver Declaration on Human Settlements, the product of the first United Nations (UN) Conference on Human Settlements (Habitat I), did not yet use the word 'slum' or anything comparable: it was focused on the rapid growth of under-resourced informal settlements and listed among its problems 'social, economic, ecological and environmental deterioration' and 'uncontrolled urbanization'. The famous 1987 Brundtland Commission report *Our Common Future* featured an urban chapter focusing on environmental problems and the problems of the urban poor in 'Third World Cities': agricultural and industrial pollution, water and sanitation, very fast growth and inadequate infrastructure, housing, and services (Brundtland 1987, 201). Habitat II in 1996 'focused essentially on managing urbanization in the global south and on the urban poor' (Parnell 2016, 532). The 2003 UN-Habitat report, *The Challenge of Slums*, highlighted dangerous and unhealthy environmental conditions in informal communities (UN-Habitat 2003).

Through the beginning of the twenty-first century, the main strategies for coping with informal settlements were to clear them or modernise them (Mayo *et al.* 1986), in both cases in the name of 'development'. But even as informal settlements were being widely cited as environmental problems and international development touted as a solution, a second narrative – of informal settlements as vibrant, innovative forms of a potentially sustainable urbanism – was growing. The year 2010 marked something of a watershed. In contrast to some of its earlier work, UN-Habitat's (2010) report *Cities for All: Bridging the Urban Divide* described urbanisation as a 'positive force for transformation' in the Global South and noted that 'too many countries have adopted an ambivalent or hostile attitude to the urbanization process, with negative consequences' (UN-Habitat's 2010, 26). That same year, Stewart Brand (2010), the American environmental visionary and founder of the Whole Earth Catalogue, wrote an article entitled 'How Slums Can Save the Planet', in which he described informal settlements as 'unexpectedly green' by virtue of their extreme density. Brand observed a 'reversal of opinion about fast-growing cities – from bad news to good news' for the environment – across the first decade of the twenty-first century. Researchers conducting interviews in informal settlements started to notice their unexpected positive qualities: their efficiency, walkability, recycling practices, economies of scale, thriving informal economies, strong networks of community support, and residents' steadily improving quality of life (Brand 2010, 31, 35–44).

Earlier scholarly work made similar points, but sporadically and without widespread popular uptake. Particularly notable was Gilbert *et al.*'s (1996) *Making Cities Work*, which discussed in detail how Northern cities could learn environmental lessons from informal settlements in the South and concluded:

The goal of sustainable development should be as much to make the North like the South as it is to make the South like the North; indeed, protection of the global environment may require that the North be made more like the South than vice versa.

(Gilbert *et al.* 1996, 14)

But now, in response to widespread clearance and displacement, informal settlements such as Dharavi in Mumbai are being upheld as hotbeds of creativity and ingenuity (Roy 2011). And many of the very characteristics that were conceptualised as environmental problems in the twentieth century – informality, impermanence, living with less, and extreme density – are now being taken up as 'models for sustainable living' (Ross 2014) as sustainability planners and designers 'take lessons from slums' (Smedley 2013). In 2016, Chilean architect Alejandro Aravena won architecture's prestigious Pritzker Prize for modular housing based on informal settlements' design and incremental growth. In the same time period, critical scholarship recognising the ingenuity of residents has also flourished. While condemning the political economy that has produced them in the first place, scholars have nevertheless elevated informal settlements as important vantage points from which to theorise global modernity (Davis 2006), as spaces of contested governance and citizenship (Appadurai 1996; Holston 2008), and of forms of distinct 'subaltern urbanism' (Roy 2011), and urban practices and culture (McFarlane 2008; Simone 2004). In short, whereas informal settlements in the Global South were once understood as environmental problems for which 'development' was the solution, they are now seen as potential resources for sustainable urbanism on a global scale.

Climate change

In the last several decades climate change has decisively emerged as the overarching environmental concern spanning the Global North and South, providing, as Matt Slavin (2011) has argued, the context for the global institutionalisation of urban sustainability. In 1987, the Brundtland Report argued that Northern cities at least in theory had the 'flexibility, space for manoeuvre, and innovation by local leadership' to deal with sustainability problems at an urban scale, unlike under-resourced Southern cities. But the report devoted more time to describing existing Northern cities' disproportionate resource use, energy consumption, environmental pollution, and internal decay (Brundtland 1987, 202–203). In the 1980s and 1990s, as scientific understandings of climate change developed, awareness of cities as sustainability problems grew. Cities were understood to be a key source of global GHG emissions, both due to the growing percentage of the world's population living in them and the lifestyle and consumption practices of their more affluent residents.

The first formal international climate change treaty effort was the 1992 UNFCCC (United Nations Framework Convention on Climate Change), which established non-binding recommendations for limiting GHG emissions and the framework for the annual COP (Conferences of the Parties) meetings that have continued since. The second was the 1997 Kyoto Protocol, which established legally binding recommendations but was plagued by questions about the role of developing countries in bearing responsibilities for climate change and the scale of reductions in industrialised ones. Kyoto's eventually agreed-upon targets were 'purely political' compromises to get the European Union, United States, and Japan on board, and they had little to do with the reductions actually necessary to stabilise CO_2 levels (Bulkeley and Betsill 2003, 32, 37–38). US President George W. Bush's withdrawal of the US from the Kyoto Protocol in 2001 did not prevent its eventual ratification but underscored the challenges and volatility of these efforts. Since then, international agreements have continued to be fraught, with US President Donald Trump's withdrawal of the US from the 2015 Paris Agreement the most recent notable breakdown.

As international efforts floundered, affluent Northern cities were increasingly cast as a great hope because of their nimble leadership, more direct accountability to constituents, and feasible scale of action. According to Bulkeley and Betsill (2003, 22), it was at the 1992 Rio Conference that cities were 'fully recognized as an area through which sustainability could and should be pursued'. They cite Gilbert *et al.* (1996, 69), who argued that Rio 'had two important consequences for the role of cities': it highlighted 'the potential role of cities in dealing with environmental issues', and 'emphasized the direct link between action on environmental issues and international cooperation between cities' (Bulkeley and Betsill 2003, 22–23). In 1996, Habitat II – 'The City Summit' – asked how to 'achieve sustainable human settlements in an urbanizing world' (Girardet 1996). Publications in the late 1990s focused on ways to help cities realise their sustainable potential through case studies that assessed the sustainability of cities – North and South – in terms of their inequality, morphology, and resource use (Gilbert *et al.* 1996) and by offering ways to assess and improve 'the environmental performance of cities in regard to the meeting of sustainable development goals' (Satterthwaite 1997). Because the majority of the world would soon live in cities, scholars and policy-makers argued, it was in and through cities that sustainability goals must be pursued. For the most part, recommendations focused on optimising urban resource use and reducing GHGs in ways that dovetailed with conversations about urban density discussed above.

By the late 1990s, the rise of the urban scale as an important target scale for governance, investment, policy, and economic development also contributed to the idea that the urban might also be a scale at which to address environmental problems. Key pieces of legislation outlining this city-based environmental agenda were the International Council for Local Environmental Initiatives (1990) and the Mayors' Climate Protection Agreement (2007). In the last several years, mayors have remained high-profile and apparently successful climate actors, as represented in projects such as the C40 Cities Climate Leadership Group, 100 Resilient Cities, and the Global Covenant of Mayors, and acts such as the Chicago Climate Charter – an agreement among mayors in Canada,

Mexico, and the United States to uphold Paris Agreement targets even after Trump announced the United States' withdrawal.

The 2000s have also been marked by a growing awareness of cities' vulnerability to extreme weather events and the rise of urban 'resiliency' planning and discourse. Through the early 2000s, climate change classics such as Weart's (2008 [2003]) *The Discovery of Global Warming* and Kolbert's (2015 [2006]) *Field Notes from a Catastrophe* barely mentioned cities, either as sites of potentially catastrophic loss of life and investment or as potential futures. Kolbert has a chapter on Dutch cities' 'floating houses' as a way of coping with sea level rise, and another on energy use in Vermont, but most illustrations of the impact of climate change on the planet focus on ecological changes in wilderness areas, such as melting ice sheets or species extinction. But weather events such as Hurricane Katrina in 2005, Hurricane Sandy in 2012, and Hurricane Harvey in 2017 – together causing hundreds of fatalities and billions of dollars in damages – raised awareness of the vulnerability of coastal cities' populations and assets to climate change. (In the same years, severe droughts in São Paulo and Cape Town, fires and mudslides in California, and snow in Europe all underscored these points.) In response, city governments and private foundations have invested increasing amounts of public and private funding in retrofitting cities for climate change. Organisations such as New York's Rebuild By Design, which began as a design competition launched by the US Department of Housing and Urban Development after Hurricane Sandy, as well as new forms of private consulting such as the exportation and circulation of Dutch water expertise, all target large cities. Today, the profile of climate change issues is significantly urban; city leaders recognise 'both unusual vulnerability and significant responsibility' for climate change impacts' (Toly 2008, 348).

From non-city to city solutions

Table 11.1 illustrates the transition from *non-city* to *city* solutions, which can be observed in each of the three environmental problems discussed above. The emergence of these urban environmental problems was not simultaneous; sprawl, slums, and climate change emerged as successive environmental problems between the 1960s and the 2000s. And each of these problems eventually led to distinctly different types of solutions in spatial planning, architecture and design, and urban governance. But each of the three policy areas underwent a similar shift. Initial efforts were *non-city* – which is to say that initial proposed solutions to problems of sprawl, slums, and climate change (limiting growth, sustainable international development, and multilateral treaty processes) did not specifically target *city* planning, *city* design, or *city* governance. Only later did *city* solutions to sprawl, slums, and climate change become dominant – in the form of compact/ smart cities, 'resilient' neighbourhoods, and strong mayoral networks – along with a distinctly urban vision of contemporary sustainability more generally.

This history illustrates the fact that the city as a solution to sustainability problems was neither necessary nor inevitable. The sustainability conversation

Table 11.1 How cities are called upon to save the planet

Global sustainability problem	Timeframe for problem emergence	Non-city solution	City solution	Type of solution
SPRAWL Too much human resource consumption, as exemplified by suburban sprawl	1960s	Limiting capitalist growth (Meadows et al. 1972); limiting urban growth (Callenbach 2009 [1975])	Concentrate people and resources in efficient urban spaces: compact cities (Jenks et al. 1996); densification (Owen 2004; Glaeser 2011); smart cities (Global Commission on Economy and Climate 2015)	Spatial planning (solution for affluent Northern cities)
SLUMS Physical vulnerability of population; sanitation, public health, extreme weather problems; too-fast growth; lack of infrastructure and services	1980s	Sustainable development (Brundtland 1987; Habitat I)	Slums as model for resilience (Brand 2010) and sustainable design (Ross 2014, Smedley 2013) Sustainable urban development (Habitat II; Girardet 1996; Gilbert et al. 1996)	Architecture/ design (solution for poor Southern cities; eventually a model for Northern cities)
CLIMATE CHANGE The need for global governance coordination to reduce GHG emissions	1990s	Multilateral treaty process (UNFCCC 1992; Kyoto Protocol 1997)	Interlocal environmental networks (ICLEI 1990); empowered mayors (Barber 2013; Mayors' Climate Protection Agreement 2007)	Governance (everywhere)
Physical vulnerability of cities' infrastructure, populations, and assets to extreme weather events	2000s	None	New investment in resilience in coastal cities (100 Resilient Cities; Rebuild By Design)	

wasn't always urban – it *became* urban. Sustainability's urban turn was a consequence, first of all, of material pressure from the growth of cities, initially in the form of concerns about sprawl and slums (which were also problems of past planning efforts) and subsequently climate change. Concerns about climate change were understood as a new kind of material pressure resulting from urban growth: worries about the vulnerability of urban assets and populations to extreme weather events, increasing global emissions as a result of accelerating urban development in the Global South, particularly China, and large carbon footprints due to the lifestyle and consumption practices of the world's more affluent urban residents.

Similarly, urban sustainability discourse need not have necessarily been 'planet-saving' in its ambitions. Sustainable cities' global problem-solving mandate reflects in part the fact that climate change is a global problem – sustainability really cannot be achieved 'in one place' (Mössner and Miller 2015) – but also the historical trajectory of sustainability as a policy issue. Contemporary sustainability policy thinking emerged out of an existing international development conversation that was already global in scope. Initial conversations about climate change were led by the United Nations, consisted of a global set of actors and social networks, and were explicitly focused on problems in both the Global North and South. The UN documents cited above are thus not simply a convenient way to trace the history of the rise of the 'cities saving the planet' discourse, but are actually part of the social, institutional, and intellectual framework out of which this common sense emerged. We would not go so far as to argue that international development efforts simply 'turned into' global urban sustainability discourse, but it is noteworthy that this non-urban set of actors, social networks, and decision-making and funding processes existed and could turn its attention to cities as sustainability solutions.

Finally, one significant difference between slums/sprawl and climate change as sustainability problems is that the first two were initially to be solved with 'less city' (by limiting growth in the North and supporting sustainable development in the South), while climate change has (thus far) offered no obvious 'less city' solution. Retreating from affluent coastal cities would have involved an unthinkable loss of assets, while by the 2000s the 'urban age' ideology (Brenner and Schmid 2015) was already consolidated. 'Less city' – or anti-urban – visions of global sustainability do exist, for instance in the form of a 'back to the land' ethos of small-scale living, but for the most part they are paths not taken in international sustainability planning and policy efforts – a counterfactual world in which Stewart Brand's *Whole Earth Catalogue* had turned California into Ecotopia instead of Silicon Valley. Instead, in the past decade, cities have become the dominant solutions the world's sustainability problems, along with a new kind of pressure to design 'better' – more resilient and sustainable – cities in the future.

A contemporary research agenda on cities saving the planet

What would an agenda for critical urban environmental research investigating the emergence of the city as a solution to environmental problems rather than the source of those problems look like? In our view, four features are central.

(1) *Historical*, emphasising the embeddedness of specific urban-environmental concerns within specific historical junctures and the path dependency of urban-environmental problems and solutions.
(2) *Multi-spatial*, emphasising the co-constitution of urban-environmental concerns across city/non-city boundaries, the multiple (sometimes conflicting) geographical scales at which urban-environmental concerns are articulated, and the uneven spatial development of urban nature.
(3) *Political*, emphasising the power differentials and conflicting interests that characterise actually-existing urban sustainability questions and the centrality of both growth and equity questions to environmental concerns.
(4) *Representational*, emphasising the importance of cultural, aesthetic, and ideological framings of cities and nature in the formation of urban sustainability policy 'common sense'.

We now proceed to elaborate these features along with some of their implications and contextualise them with respect to existing scholarship.

(1) Historical

To begin with, contextualising present trends in urban sustainability policy and politics – including the broad notion that cities can save the planet – in historical terms reveals both important continuities and ruptures. The historical account of the emergence of the 'cities can save the planet' paradigm offered in the previous section suggests that *different modes of urban development facilitate different framings of environmental problems and solutions*. The land-intensive mode of suburban expansion, which characterised the post-war Global North, encouraged a set of concerns related to unsustainable resource and land consumption. And the rapid expansion of informal settlements in the Global South from the 1970s onward encouraged a set of concerns related to the environmental deficits of unplanned growth. Extending this observation into the present, technology-led urban development in post-industrial cities should likewise be expected to yield its own characteristic set of environmental concerns. At the same time, an emphasis on the historical embeddedness of particular configurations of the urban–environmental nexus is a potentially powerful corrective to ahistorical thinking, which sees the return of nature to the city as a uniquely contemporary development. By contrast, an historical approach to urban nature reveals what is new about contemporary urban sustainability thinking but also what has endured – including concerns about the socially induced disappearance of nature, environmentalism as an accumulation strategy, and patterns of racial exclusion. An historical approach finally allows us to identify counterfactuals: other possibilities for how urban environments and urban environmentalisms might have developed and which still might serve as resources for the present.

(2) Multi-spatial

Urban studies has increasingly embraced a vision of urbanisation as a multi-spatial process which exceeds the traditional concept of the city (Wachsmuth 2014). However, and in spite of their self-consciously global scope, *dominant forms of urban sustainability planning and thinking focus too narrowly on cities.* Planetary urbanisation (Brenner and Schmid 2015) is not simply a world of cities, and the global dimensions of urban environmental processes are likewise not simply nature in the city (Angelo and Wachsmuth 2015). Accordingly, one key issue for urban sustainability research to address is the relationship between cities and other place-distinctions within the urban fabric, such as rural, suburban, and peri-urban spaces. For example, Mössner and Miller (2015) document the regional contradictions of local sustainability policy in a comparative analysis of Freiburg, Germany and Calgary, Canada – where the central city's 'sustainability fix' is met with a suburban 'counter-sustainability fix'. Environmental interconnections between localities also take more distantiated form, and the development of global/local networks of urban sustainability policy expertise is also an urgent topic for research. Finally, further afield from the city or even the global city network, in the absence of a multi-spatial framework, 'operational landscapes' (Brenner and Katsikis 2014) of energy, resource extraction, logistics, and waste processing, which support urban agglomerations, risk being rendered invisible (Arboleda 2016).

(3) Political

Contemporary urban sustainability research should entail a recognition of – and challenge to – the fact that dominant forms of urban sustainability planning and thinking are socially narrow, often focusing on the livability concerns of wealthy urban residents to the exclusion of the poorer residents of urban peripheries (Wachsmuth *et al.* 2016). More broadly, urban sustainability policy has developed in a pro-growth, system-affirmative direction that stands in rather stark contrast to its explicitly or implicitly anti-capitalist antecedents from the post-war era. While it would be a mistake to posit too clean a distinction between earlier, radical forms of sustainability thinking and the present embrace of sustainability by the 'power elite' (Greenberg 2015), the fact remains that an earlier system of environmental thought premised on concepts such as the 'limits to growth' (Meadows *et al.* 1972) has been largely supplanted by a system of thought premised on sustainability as the sustainability of growth. Given the present centrality of urbanisation to global capitalist development, interrogating the contradictions of urbanising green capitalism is an important and promising line of research (Knuth 2017). As Greenberg (2015, 108) argues, 'insofar as a market-oriented discourse of sustainability becomes a dominant and powerful agent within contemporary capitalism and capitalist urbanization, it has the capacity to render other, nonmarket goals – whether ecological or social –unsustainable'.

(4) Representational

Lastly, an adequate research agenda into the politics of cities saving the planet should be attentive to the different representations and ideologies that structure and constrain these politics. As we have explored elsewhere (Wachsmuth and Angelo 2018), contemporary urban sustainability thinking is strongly characterised by a pair of opposing but also complementary representations of the city–nature relationship: 'green urban nature' and 'grey urban nature'. These aesthetic bundles of associations between what is urban and what is sustainable inform implicit but consequential policy common senses and imaginaries. Other, non-aesthetic dimensions of urban sustainability representation also structure how environmental policymaking common senses are formed. The narrow set of 'global cities' which supply a disproportionate share of global best practices are a case in point – specific privileged ideas of urban environments which circulate alongside the policies themselves. Indeed, the very notion of 'cities saving the planet' is arguably a subset of a broader political and ideological yearning for 'mayors ruling the world' (Barber 2013). Localities and their political leaders are valorised as nimble, pragmatic actors in contrast to supposedly ineffective and obsolete nation-states, and current developments in the realm of sustainability thinking are to some extent a rehashing of earlier political and economic debates. In the 1990s, economic globalisation sparked emphatic pronouncements by scholars and policy-makers about the declining relevance of national borders and nation-states (Ohmae 1996), with the result that the local became understood 'as a site of empowerment in the new global age' (Brenner and Theodore 2002, 342), and a 'new localism' was proposed as a political strategy (Goetz and Clarke 1993). Critical correctives explained new global/local dynamics as forms of state 'reterritorialization' (Brenner 1999) and 'glocalization' (Swyngedouw 1997) rather than a diminishment of place, nation state, or territorial identity in absolute terms, but those insights have arguably been underdeveloped in urban sustainability research (although see Marvin and Guy 1997).

Conclusion

In its dominant and most visible forms, contemporary sustainability policy is markedly city-centric and system-affirmative. Especially in the context of present concerns regarding climate change and growing social inequality, it is interesting to note that early strands of urban sustainability discourse exhibited seeds of concern for other issues that were never really taken up. Early thinking on cities as potentially sustainable environments made arguments about desirable urban sustainability agendas which still read as cutting edge. Gilbert *et al.*'s (1996) *Making Cities Work*, for instance, insisted that inequality and climate change – 'protecting the environment and combating poverty' – were 'interlinked issues' that had to be solved together (Gilbert *et al.* 1996, 12) and in a spatial frame that took both cities and their hinterland and the place of cities in the world into account (see also Satterthwaite 1997). This work underscores the fact that there were other possible directions in which sustainability policy could have headed. It may be worth revisiting these paths not taken.

The purpose of this chapter has been to identify and explain the development of a new frame for understanding the relationship between cities and the environment on the global city, summed up in the question 'why does everybody thinks cities can save the planet?' Sustainability and urbanism have become a powerful pair of master discourses for our time, and their intersection suggests itself as a solution for a wide set of social, economic and environmental problems. We have narrated the emergence of the 'cities saving the planet' frame through the historical development of three successive juxtapositions of cities and environmental problems – suburban sprawl, the proliferation of slums, and challenge of climate change – and indicated how these juxtapositions collectively demonstrate a movement from cities being understood as environmental problems towards cities being understood as the *solution* to environmental problems on a global scale. On the basis of this discussion, we have elaborated a research agenda for investigating cities saving the planet based on the four principles of historical, multi-spatial, political, and representational analysis.

Note

* This chapter is reproduced with permission from SAGE.

References

Ahern, J. (2011) From fail-safe to safe-to-fail: sustainability and resilience in the new urban world. *Landscape and Urban Planning*, 100(4), 341–343.

Angelo, H. and Vormann, B. (2018) Long waves of urban reform: putting the smart city in its place. *City*, 1–19.

Angelo, H. and Wachsmuth, D. (2015) Urbanizing urban political ecology: a critique of methodological cityism. *International Journal of Urban and Regional Research*, 39(1), 16–27.

Appadurai, A. (1996) *Modernity at Large: Cultural Dimensions of Globalization*, Vol. 1. Minneapolis, MN: University of Minnesota Press.

Arboleda, M. (2016) Spaces of extraction, metropolitan explosions: planetary urbanization and the commodity boom in Latin America. *International Journal of Urban and Regional Research*, 40(1), 96–112.

Barber, B.R. (2013) *If Mayors Ruled the World: Dysfunctional Nations, Rising Cities*. New Haven, CT: Yale University Press.

Brand, S. (2010) How slums can save the planet. *Prospect Magazine*.

Breheny, Michael J. (ed.) (1992) *Sustainable Development and Urban Form*, Vol. 2. London: Pion.

Brenner, N. (1999) Globalisation as reterritorialisation: the re-scaling of urban governance in the European Union. *Urban studies*, 36(3), 431–451.

Brenner, N. (1999) Beyond state-centrism? Space, territoriality, and geographical scale in globalization studies. *Theory and Society*, 28(1), 39–78.

Brenner, N. and Katsikis, N. (2014) Is the Mediterranean urban. *Implosions/Explosions: Towards a Study of Planetary Urbanization*. Berlin: Jovis, 428–459.

Brenner, N. and Schmid, C. (2015) Towards a new epistemology of the urban? *City*, 19(2–3), 151–182.

Brenner, N. and Theodore, N. (2002) Preface: from the 'new localism' to the spaces of neoliberalism. *Antipode*, 34(3), 341–347.

Bruegmann, R. (2006) *Sprawl: A Compact History*. Chicago, IL: University of Chicago Press.

Brundtland, G.H. (1987) *Report of the World Commission on Environment and Development:' Our Common Future'*. New York: United Nations.

Bulkeley, H. and Betsill, M. (2003) *Cities and Climate Change: Urban Sustainability and Global Environmental Governance*. New York: Routledge.

Burchell, R.W., Listokin, D., and Galley, C.C. (2000) Smart growth: more than a ghost of urban policy past, less than a bold new horizon. *Housing Policy Debate*, 11(4), 821–879.

Burgess, R. and Jenks, M. (eds) (2002) *Compact Cities: Sustainable Urban Forms for Developing Countries*. New York: Routledge.

Callenbach, E. (2009[1975]) *Ecotopia*. New York: Bantam.

Calthorpe, P. (1985) Redefining cities. *Whole Earth Review*, 45, 1–3.

Carson, R. (2002 [1962]) *Silent Spring*. Cambridge, MA: Houghton Mifflin Harcourt.

Commoner, B. (1971) *The Closing Circle: Nature, Man, and Technology*. New York: Random House.

Davis, M. (2006) *Planet of Slums*. London and New York: Verso.

Duany, A., Speck, J., and Lydon, M. (2010) *The Smart Growth Manual*. New York: McGraw-Hill.

Du Pisani, J.A. (2006) Sustainable development – historical roots of the concept. *Environmental Sciences*, 3(2), 83–96.

Florida, R.L. and Feldman, M. (1988) Housing in US Fordism. *International Journal of Urban and Regional Research*, 12(2), 187–210.

Gilbert, R., Stevenson, D., Girardet, H., and Stren, R. (1996) *Making Cities Work*. London: Earthscan.

Girardet, H. (1996) *The Gaia Atlas of Cities: New Directions for Sustainable Urban Living*. New York: UN-HABITAT.

Girardet, H. (1999) *Creating Sustainable Cities*. Devon: Green Books.

Girardet, H. (2007) Creating livable and sustainable cities. *Surviving the Century: Facing Climate Chaos and Other Global Challenges*. Sterling, VA: Earthscan, 103–126.

Glaeser, E. (2011) *Triumph of the City: How Urban Spaces make us Human*. Hampshire: Pan Macmillan.

Global Commission on Economy and Climate (2015). *Seizing the Global Opportunity*. Washington, DC: New Climate Economy.

Goetz, E.G. and Clarke, S.E. (eds) (1993) *The New Localism: Comparative Urban Politics in a Global Era*, Vol. 164. Newbury Park, CA: Sage Publications.

Greenberg, M. (2015) 'The sustainability edge': competition, crisis, and the rise of green urban branding. In C. Isenhour, G. McDonogh, and M. Checker (eds), *Sustainability in the Global City: Myth and Practice*. Cambridge, UK: Cambridge University Press, 105–130.

Hamel, P. and Keil, R. (eds) (2015) *Suburban Governance: A Global View*. Toronto: University of Toronto Press.

Hollands, R.G. (2008) Will the real smart city please stand up? *City*, 12(3), 303–320.

Holston, J. (2008) *Insurgent Citizenship: Disjunctions of Democracy and Modernity in Brazil*. Princeton, NJ: Princeton University Press.

Jenks, M., Burton, E., and Williams, K. (1996) *The Compact City. A Sustainable Urban Form*. London: E. & F.N. Spon.

Keil, R. (ed.) (2013) *Suburban Constellations: Governance, Land and Infrastructure in the 21st Century*. Berlin: Jovis Verlag.

Knuth, S. (2017) Green devaluation: disruption, divestment, and decommodification for a green economy. *Capitalism Nature Socialism*, 28(1), 98–117.

Kolbert, E. (2015 [2006]) *Field Notes from a Catastrophe: Man, Nature, and Climate Change*. New York: Bloomsbury Publishing USA.

MacKinnon, D. and Derickson, K.D. (2013) From resilience to resourcefulness. *Progress in Human Geography*, 37(2), 253–270.

Marvin, S. and Guy, S. (1997) Creating myths rather than sustainability: the transition fallacies of the new localism. *Local Environment*, 2(3), 311–318.

Mayo, S.K., Malpezzi, S., and Gross, D.J. (1986) Shelter strategies for the urban poor in developing countries. *The World Bank Research Observer*, 1(2), 183–203.

McFarlane, C. (2008) Governing the contaminated city: infrastructure and sanitation in colonial and post-colonial Bombay. *International Journal of Urban and Regional Research*, 32(2), 415–435.

Meadows, D.H., Meadows, D.L., Randers, J., and Behrens, W.W. (1972) *The Limits to Growth: A Report for the Club of Rome's Project on the Predicament of Mankind*. New York: New American Library.

Mössner, S. and Miller, B. (2015) Sustainability in one place? Dilemmas of sustainability governance in the Freiburg metropolitan region. *Regions Magazine*, 300(1), 18–20.

Newman, P.G. and Kenworthy, J.R. (1989) *Cities and Automobile Dependence: An International Sourcebook*. Aldershot, UK: Gower.

Ohmae, K. (1996) *End of the Nation State: The Rise of Regional Economies*. New York: Harper Collins Publishers.

Owen, D. (2004) Green Manhattan. *The New Yorker*, 80(31), 111–123.

Parnell, S. (2016) Defining a global urban development agenda. *World Development*, 78, 529–540.

Pulido, L. (2000) Rethinking environmental racism: white privilege and urban development in Southern California. *Annals of the Association of American Geographers*, 90(1), 12–40.

Rome, A. (2001) *The Bulldozer in the Countryside: Suburban Sprawl and the Rise of American Environmentalism*. Cambridge, UK: Cambridge University Press.

Ross, P. (2014) Climate change solutions: architects look to slums as models for sustainable living. *International Business Times*. Available at: www.ibtimes.com/climate-change-solutions-architects-look-slums-models-sustainable-living-1623418 (accessed May 2018).

Roy, A. (2011) Slumdog cities: rethinking subaltern urbanism. *International Journal of Urban and Regional Research*, 35(2), 223–238.

Satterthwaite, D. (1997) Sustainable cities or cities that contribute to sustainable development? *Urban Studies*, 34(10), 1667–1691.

Satterthwaite, D. (2016) Successful, safe and sustainable cities: towards a New Urban Agenda. *Commonwealth Journal of Local Governance*, 3–18.

Schneider, A. and Woodcock, C.E. (2008) Compact, dispersed, fragmented, extensive? A comparison of urban growth in twenty-five global cities using remotely sensed data, pattern metrics and census information. *Urban Studies*, 45(3), 659–692.

Sellers, C.C. (2012) *Crabgrass Crucible: Suburban Nature and the Rise of Environmentalism in Twentieth-century America*. Chapel Hill, NC: University of North Carolina Press.

Sheller, M. and Urry, J. (2000) The city and the car. *International Journal of Urban and Regional Research*, 24(4), 737–757.

Sierra Club (1998) 1998 sprawl report. Available at: http://vault.sierraclub.org/sprawl/report98/report.asp.

Simone, A. (2004) *For the City Yet to Come*. Durham, NC: Duke University Press.

Slavin, M.I. (2011) The rise of the urban sustainability movement in America. In *Sustainability in America's Cities*. Washington, DC: Island Press, 1–19.

Smedley, T. (2013) Sustainable urban design: lessons to be taken from slums. *Guardian*. Available at: www.theguardian.com/sustainable-business/sustainable-design-lessons-from-slums (accessed May 2018).

Swyngedouw, E. (1997) Neither global nor local: 'glocalization' and the politics of scale. In K. Cox (ed.), *Space of Globalization: Reasserting the Power of the Local*. New York/London: Guilford/Longman 115–136.

Toly, N.J. (2008) Transnational municipal networks in climate politics: from global governance to global politics. *Globalizations*, 5(3), 341–356.

UN-Habitat (2003) *The Challenge of Slums: Global Report on Human Settlements 2003*. London: Earthscan.

UN-Habitat (2010) *State of the World's Cities 2010/11: Cities for All, Bridging the Urban Divide*. New York: UN-Habitat.

UN-Habitat (2012) *State of the World's Cities 2008/9: Harmonious Cities*. New York: Routledge.

Wachsmuth, D. (2014) City as ideology: reconciling the explosion of the city form with the tenacity of the city concept. *Environment and Planning D: Society and Space*, 32(1), 75–90.

Wachsmuth, D. and Angelo, H. (2018) Green and gray: new ideologies of nature in urban sustainability policy. *Annals of the American Association of Geographers*, 108(4), 1038–1056.

Wachsmuth, D. Cohen, D.A., and Angelo, H. (2016) Expand the frontiers of urban sustainability. *Nature News*, 536(7617), 391.

Walker, R. (1981) *Theory of Suburbanization: Capitalism and the Construction of the Urban Space in the United States*. London: Methuen, 383–429.

Weart, S.R. (2008 [2003]) *The Discovery of Global Warming*. Cambridge, MA: Harvard University Press.

Wheeler, S.M. (2000) Planning for metropolitan sustainability. *Journal of Planning Education and Research*, 20(20), 133–145.

12 Imagining the net zero emissions city

Urban climate governance in the City of Melbourne, Australia

Stephen Pollard

Introduction

The target of net zero emissions has emerged as an ideal and imperative response to climate change across multiple levels of climate governance, including local government. This chapter examines political commitments and pragmatic shifts in the City of Melbourne's successive strategies to become a net zero emissions city. The launch of *Zero Net Emissions by 2020* (ZNE2020) in 2003 marked the City of Melbourne as one of the first local governments anywhere to commit to 'community-wide' carbon neutrality (City of Melbourne 2003). Periodic updates to the strategy saw the City's initial emphasis on carbon trading switch towards accelerating the transition from fossil fuels to renewable energy. With the 2020 timeframe drawing near, the City's *Climate Change Mitigation Strategy to 2050*, released in December 2018, maintains the political commitment to net zero emissions but reframes the policy narratives, timeframes, geographies, technologies, and scales around this objective. Charting these phases against a backdrop of shifting circumstances – political, technological, and environmental – unveils persistent tensions between local government ambition and authority to act and between conceptions of the net zero emissions city as locally discrete and globally networked.

In this chapter, I explore how the City's successive imaginaries of the net zero emissions city and the mechanisms designed to achieve it are framed as both politically imperative and socially and economically beneficial. The concept of 'sociotechnical imaginaries' (Jasanoff and Kim 2015) is used to trace envisaged and actual reconfigurations of urban socio-technological systems and the forms of knowledge and power that shape these as viable and legitimate within local contexts and wider networks. Underlying urban green transitions are questions of whom the imagined city of net zero emissions is for and who is doing the imagining. Cities are not self-contained: they rely on extensive systems of sociotechnical infrastructure and resource flows to enable their reproduction, and notions of urban sustainability are inevitably tied up with wider issues of social and ecological unevenness within and beyond cities (Marvin and Hodson 2010). Establishing commitments to the net zero emissions city relies on delineating conceptual boundaries around the municipality in order to balance carbon across those edges (Kenis and Lievens 2017). These interfaces are often ambiguous and involve frictions between mundane and ubiquitous urban activities, technical knowledge of carbon, and wider networks of socio-technical infrastructures and resource flows.

The analysis is based on documents related to ZNE2020 and updates to this strategy in 2008 (City of Melbourne 2008) and 2014 (City of Melbourne 2014) through to the City of Melbourne's *Climate Change Mitigation Strategy to 2050* (City of Melbourne 2018). Analysis of discourses contained in these documents is guided and supported by research interviews with a range of key actors involved in the City's climate policies at various stages. The analysis is focused on the City's mitigation strategies with particular attention paid to possibilities and practices of balancing carbon across the conceptual edges of the city. The Australian context provides an opportunity to examine urban climate govern-ance amidst a backdrop of uncertainty and upheaval over climate and energy policy. Tracing the City of Melbourne's efforts to maintain its political commit-ment to the net zero emissions city while responding to this turbulence reveals instances of making, and remaking, conceptual boundaries around the muni-cipality with respect to carbon. Exploring how these boundaries are asserted and maintained over time, and with whom, offers insights into complex dynamics of sociotechnical change in urban green transitions. Throughout this text, 'City of Melbourne' and 'City' refer to the city administration, while 'city' refers to the municipal area. While this analysis offers a critical perspective, my overall aim is to acknowledge the City of Melbourne's ongoing commitment and invention in responding to climate change.

The chapter is structured as follows. First, I introduce concepts of net zero emissions, municipal carbon accounting, and sociotechnical imaginaries in rela-tion to processes of boundary-making in urban climate governance. These con-cepts are used to explore ways in which climate threats and accountabilities are re-scaled between global and local levels and how locally bounded targets for net zero emissions are established as legitimate and beneficial. Next, I describe some key characteristics of the City of Melbourne and then turn to a sequential analysis of the ZNE2020 strategy and its 2008 and 2014 updates. This chron-ology charts the City's shifting preferences from developing a local market in carbon trading to investing in renewable energy. Underneath these shifts, the analysis explores ongoing tensions across municipal boundaries. These relate to the ambition of the City's net zero emissions imaginary and its reliance on actors outside the municipality fulfil this ambition via carbon trading and renewable energy to reduce and offset emissions within those boundaries. Turning last to the 2018 climate strategy, these tensions are restrung as the City of Melbourne reframes the temporal and spatial boundaries of its net zero emis-sions target, and in doing so repositions itself in relation to neighbouring metro-politan municipalities, the Victorian State Government, and other world cities through the C40 Cities network. In conclusion, the analysis demonstrates the situated and contingent nature of the City of Melbourne's political commit-ments to the net zero emissions city. Asserting, stabilising, and maintaining the conceptual boundaries around this target involves ongoing tensions concerning limits to local government authority and actions to manage carbon within and outside of local areas.

Boundary-making in urban climate governance

Net zero emissions and urban climate governance

The overriding objective of the international 2015 Paris Agreement on climate change is to achieve global net zero emissions – a balance of greenhouse gas (GHG) emissions sources and sinks – by the second half of the century in order to avoid dangerous climate change (UNFCCC 2015). While the concept of net zero emissions appears straightforward enough as a technical measurement, the assessment of GHG sources and sinks towards this end is complex. There are substantial difficulties involved in deciding what a 'safe' level of climate change is and for whom, what this end state will comprise, how to get there, and even where we are starting from (Geden and Beck 2014; Knutti *et al.* 2016; Schleussner *et al.* 2016). Such uncertainties and indeterminacies are not a reason to doubt the truth or seriousness of climate change. Rather, they highlight the complexity of issues at the intersection of scientific knowledge, science-based technologies, and risk (Jasanoff *et al.* 1998; Miller and Edwards 2001; Thompson and Rayner 1998).

Notwithstanding these ambiguities, nation states around the world have endorsed the ambition of global net zero emissions, although to date there is an underwhelming lack of firm commitments and actions through which to realise it (UNEP 2018). Coupling these imperatives with pressures of globalisation and urbanisation, cities have been identified as key sites to contribute to the Paris Agreement goals and to realise visions of desirable urban futures (Acuto 2016; Agarwala 2015; Broto 2017; Jordan *et al.* 2015). Amongst urban sustainability narratives, a growing number of local governments are setting targets and time-tables for net zero emissions across their municipal areas. These targets describe a situation where GHG emissions are reduced as much as possible within a discrete geographic area, such as municipal territory, and any emissions that remain are counteracted by reducing or capturing an equal amount of GHG emissions elsewhere. The idea has taken hold through several transnational municipal networks, which offer a powerful platform for cities, or those speaking on their behalf, to actively participate in world politics (Acuto and Rayner 2016; Betsill and Bulkeley 2006; Bulkeley *et al.* 2014). For example, the Carbon Neutral Cities Alliance represents 20 municipal governments from around the world with commitments to reduce emissions by 80–100 per cent and become carbon-neutral by 2050 (Carbon Neutral Cities Alliance). The C40 Cities network is supporting its almost 100-member cities to become carbon-neutral by 2050 under its Deadline 2020 programme (C40 Cities and Arup 2016).

Rescaling climate change through municipal carbon accounting

Fundamental to making the concept of net zero emissions operational at local scales is the technical knowledge generated through municipal carbon accounting. Municipal carbon accounting not only renders carbon visible but also rescales concerns about global climate change to local levels and frames accountabilities and responsibilities of local actors (Callon 2009; Kuch 2015; MacKenzie 2009). Following Anders Blok (2013, 6), municipal carbon accounts can be understood as a device to rescale knowledge of global ecological risks into situated city-making practices. At the same time, such accounts offer the potential to aggregate the impact of local emissions reduction efforts and rescale urban city-making practices into matters of global importance.

For municipal governments, the Global Protocol for Community-scale GHG Emissions Inventories (GPC) has emerged as the standard approach to calculating territorial emissions within discrete local areas (Greenhouse Gas Protocol 2016). Municipal carbon accounts seek to localise GHG emissions within territorial boundaries and administrative control by delineating emissions 'scopes'. These assign emissions according to where they are generated, including from direct fuel combustion (scope 1), from outside municipal boundaries such as stationary energy generated to meet local demand (scope 2), and embodied in supply chains and waste (scope 3) (Greenhouse Gas Protocol 2016). These spatial definitions are useful, but emissions also traverse complex and overlapping regimes of political authority and domains of social and economic life. Defining different system boundaries results in significant differences between calculated emissions, potentially redirecting mitigation efforts in varied ways (Davis and Caldeira 2010; Liu *et al.* 2015).

Managing carbon towards a defined end point is not only a technical exercise: it involves certain judgements about which emissions sources and sinks to include and exclude, how to value and aggregate different types of emissions, how to make comparisons over time, and how to use the information to inform policy (Callon 2009; MacKenzie 2009). Such issues play into disputes about how much cities, broadly defined, are to blame for climate change, such as which cities and what structures and agents within and beyond those cities ought to be held to account for urban emissions (Dodman 2009; Satterthwaite 2008). While cities concentrate many of the human activities, institutions, and systems of infrastructure that generate GHG emissions, their reproduction relies on extensive networks of sociotechnical infrastructure and continued flows of resources (Marvin and Hodson 2010). Thus, efforts to measure and manage urban GHG emissions involve negotiating boundaries around complex and interconnected ecological and socio-technical systems.

The net zero emissions city as a sociotechnical imaginary

Conceiving the net zero emissions city as a sociotechnical imaginary is a way to explore processes of boundary-making within urban climate governance. Political commitments and policy strategies to realise net zero emissions cities bridge local and global issues and encompass visions of desirable social and technological order that extend beyond cities. Sheila Jasanoff and Sang Hyun Kim define sociotechnical imaginaries as 'collectively imagined forms of social life and social order reflected in the design and fulfilment of nation-specific scientific and/or technological projects' (Jasanoff and Kim 2009, 120). The concept helps to draw out how collective representations of desirable futures are shaped, especially with respect to new and reconfigured sociotechnical assemblages such as systems of energy provision, transport, food production, or habitation. Such imagined futures express implicitly notions of the public good and are used to inform and legitimise policy actions. And while sociotechnical imaginaries are often thought of in relation to a sense of national political identity, they are also embedded in and produced by individual or collective accounts of potential futures across scales (Jasanoff 2015). This prompts us to ask how sociotechnical imaginaries operate across scales to shape and justify responses to climate risk, preferred pathways of social and technological change, and visions of a desirable future in the context of climate change.

By asserting the boundaries of the city as well as the flows of carbon that cross those edges, sociotechnical imaginaries of the net zero emissions city shape and reinforce political dimensions of knowledge production while at the same time reshaping ecological, social, and technological relations. Michael Hodson and Simon Marvin (2017) argue that the dominant framings of urban sustainability are based on assumptions of 'green growth', where economic growth is used to ecologically modernise urban environments. These heroic narratives have become increasingly narrowed towards elements of urban environments with economic and market potential, marginalising traditional concerns with social justice and equity within, and beyond, cities themselves (Gleeson 2014). In this context, it seems important to consider the conditions under which consent and trust are generated in imaginaries of the net zero emissions city. This begs the questions of what kinds of knowledge are used, amongst which networks and actors, and how uneven and fragile the conditions are for creating such consent and trust. This includes the extent to which local imaginaries of urban low carbon transition are shaped by, or able to influence, wider contexts of state and national policies and dominant interests in emissions-intensive sectors of energy, transport, construction, retail, and so on. To examine these processes empirically, we now turn to consider the City of Melbourne and its successive efforts to define, enact, and maintain its vision for a net zero emissions city.

Situating the City of Melbourne

The City of Melbourne covers an area of 37.7 square kilometres at the geographic centre of Greater Melbourne. It is one of 32 municipal governments within this larger metropolis of almost 10,000 square kilometres and 4.8 million people (City of Melbourne 2018, 10). At present, over 158,000 people live within the municipal area, while the daily influx of workers, students, and visitors to the central city is estimated at 928,000 (City of Melbourne 2019).

The municipality is located on the lands of the Wurundjeri people of the Kulin nation and encompasses the densely developed city centre and 11 surrounding suburbs. The central business district comprises wide boulevards laid out on a grid, interspersed with small laneways celebrated for their lively music venues, street art, and international food. Skyscrapers tower overhead and present a formidable skyline, and construction cranes display its continued expansion. Corporate headquarters and government agencies, universities and medical precincts, galleries and museums, festivals and events, position the city at the centre of the State of Victoria's productivity and innovation. The central city is fringed by the Yarra River, which is called Birrarung Marr in the local indigenous language, and is surrounded by parklands and sporting arenas, including the iconic Melbourne Cricket Ground. These give way to an assortment of urban environments, including the Port of Melbourne's heavy industry, sites of urban renewal, gentrified inner suburbs and public housing estates, shopping strips and public parks, all interspersed with networks of road, rail, footpaths, and cycling paths.

As these assorted urban forms suggest, activities in the city are diverse and varied. The City of Melbourne has authority over planning and building, rates, local roads, waste management, parks and gardens, amongst other things. However, as a former executive in the city administration explained, 'the ability of cities to influence varies dramatically. Some cities run a lot of big buildings, schools, bus fleets. The City of Melbourne is unusually small in terms of its jurisdictional authority' (Research interview, November 2017).

Much of the activity within the city that causes emissions lies outside the domain of local government authority. Emissions from stationary energy are very high due to Victoria's reliance on brown coal for electricity. The state government also powerfully shapes the urban environment through land-use zoning, planning, and building controls for large developments, and major infrastructure and services, including toll roads and public transport. Coupled with this is ambivalence over the role of government in shaping decisions and actions that relate to individual autonomy and private property. As a Councillor at the City of Melbourne said, 'it's easy enough to make sustainability decisions about the public realm, but far more challenging to influence decisions about the private realm' (Research interview, December 2018). Thus, the city administration is one actor amongst many involved in shaping the future direction of the central city and the wider Melbourne metropolis. As the following sections explore, the City of Melbourne's efforts over time to frame and realise its targets for the net zero emissions city reveals complex dynamics and tensions involved in urban green transitions.

Zero net emissions by 2020: a roadmap to a climate neutral city

In 2002, the City of Melbourne announced its commitment to reduce greenhouse gas emissions and achieve net zero emissions by 2020, both for its own operations and 'community-wide' across the municipality. It published its first strategy towards this objective in 2003, *Zero Net Emissions by 2020: A Roadmap to a Climate Neutral City*, establishing the municipality as an early mover in efforts by cities to respond to climate change. ZNE2020 frames climate change as a matter of global environmental risk requiring cities and their citizens to contribute a fair share towards collective action. In this respect, the strategy sets out to 'end the city's contribution to climate change', encompassing the local government and the commercial, industrial, and residential inhabitants of the city (City of Melbourne 2003, 3). The political commitment to alleviate the municipality's contribution to climate change was also designed to elicit support from local businesses and residents with its positive portrayal of the economic and social benefits that would flow from these actions.

Alongside its objectives on climate change, the City was advancing multiple plans to position Melbourne on the world stage. As one former Councillor explained,

> the trends of globalisation were taking hold, and there was a sense that Melbourne could occupy its own niche in this global context if it could develop the capacity to attract finance, investment, knowledge, and people from around the world.
>
> (Research interview, February 2018)

Besides assessing climate risks, the strategy framed rapid urban growth in Australian cities and overseas as threats to the urban environment and the quality of urban life, and globalisation as changing the nature of competition between cities for investment. These risks were cast as opportunities to position Melbourne as an attractive centre for 'green' productivity that coupled 'productive, knowledge-based industries with a quality lifestyle and environment' (City of Melbourne 2003, 3).

The conceptual boundaries of the net zero emissions target were defined, at least in part, by the territorial boundaries of the municipality and the jurisdictional authority of local government. The City's first municipal carbon accounts delimited an area aligned to the administrative boundary of the municipality, identifying total emissions of 3.75 million tonnes of carbon dioxide equivalent (CO_2^{-e}) across commercial buildings, industry, residents, and waste (City of Melbourne 2003, 15). Demand for stationary energy was identified as the major source of emissions from activities within the municipality. The accounts excluded transport and embodied energy, noting their small contribution to total emission and the Council's limited influence over these areas (City of Melbourne 2003, 5), although these were brought into later strategies. The strategy set out actions for the net zero emissions city in three core areas: 'leading edge design' to reduce energy demand, 'decarbonising electricity' by influencing the renewable energy generation, and 'carbon sequestration' to counteract remaining emissions with carbon offsets (City of Melbourne 2003, 4). And although corporate emissions from the City of Melbourne's own operations accounted for less than 1 per cent of total 'community wide' emissions, the City was keen to leverage its authority and leadership by reducing emissions across its assets and activities (City of Melbourne 2003, 9).

From the outset, the strategy anticipated measures to reduce emissions outside the municipal boundaries and beyond the Council's direct control, indicating limits to local government power and agency in matters affecting GHG emissions. The net zero emissions target rested in large part on the City's perception of the inevitability of global carbon markets and the emergence of policy and legal frameworks to participate in these markets. Discourses around carbon trading had been emerging through the international climate regime as a viable, low cost, and 'optimal' response to climate change (City of Melbourne 2003, 3). Conditions for such markets to emerge were being established both locally and internationally with the concept of carbon offsets (Bumpus and Liverman 2008). At the state level, the Victorian Government had amended the *Forestry Rights Act 1996* to encourage investment in forest plantations as carbon sinks by disaggregating legal rights to land, trees, and carbon and allowing carbon held in certain trees to be traded (City of Melbourne 2003, 44). Nationally, carbon trading was emerging as the preferred response to climate change, with the *Bush for Greenhouse* programme developing tools to account for the carbon content of certain types of land use and forestry, and working on legal instruments, standards, and guidance that would later provide the structure for carbon trading in Australia (Kuch 2015). These technical details were not all straightforward. A former technical adviser to this programme recounted negotiating over whether trees that kangaroos could jump over could still be counted as trees in calculations for carbon sequestration (Research interview, November 2017).

Rather than waiting for these markets to emerge from above, the City envisaged a local carbon trading market to provide a 'least cost, flexible and effective collective response' while preparing the city administration and resident businesses for 'eventual international trading in carbon credits' (City of Melbourne 2003, 3). Achieving net zero emissions would rely on establishing these flows of carbon as viable and legitimate. The 2003 strategy proposed a pilot to offset a portion of the Council's own emissions by purchasing the carbon rights for blue gum mallee eucalypts, identified as a carbon sink and a profitable investment (City of Melbourne 2003, 54). The pilot programme would see 10 per cent of corporate emissions sequestered in a plantation located in a specified municipality in the north of the state, increasing to 50 per cent by 2010. The eucalypts would store carbon in their roots, while the growth above ground would be used as feedstock for power generation, with eucalyptus oil as a valuable by-product (City of Melbourne 2003, 41). 'Combined with the 50 per cent target in the City of Melbourne's use of renewable energy, sequestration will deliver net zero corporate emissions' (City of Melbourne 2003). While this action was designed around the City's corporate emissions profile, the longer term would see the pilot developed into a carbon trading scheme for the city's businesses, residents, and industries as a way to encourage these actors to 'shoulder their own responsibility' for GHG emissions (City of Melbourne 2003, 53).

The narrative of ecological modernisation resonates strongly in this framing of individual responsibility, technological innovation, and market opportunity as a way to address environmental ills and reap economic and social benefits. If the City of Melbourne moved fast enough, the strategy suggested, it might create a competitive advantage similar to the City of London, which was 'already positioning itself as a global hub for greenhouse gas emissions trading' (City of Melbourne 2003, 15). Although the ZNE2020 strategy framed a market-based solution as logical and natural, it also placed limits on how far that solution ought to extend. The strategy asserted that the location for carbon trading 'must be in Victoria – it would be inappropriate for the City of Melbourne to partner with an organization from another State' (City of Melbourne 2003, 43). The strategy suggested that pre-existing partnerships within the municipality that were proposed to host the eucalypt plantation would help to develop trust with participants in the nascent market. It also suggested that rural and urban communities would be encouraged to 'learn of the wider impact of their simple day-to-day actions on their household, the environment and the wider community' (City of Melbourne 2003, 46). Although based in market exchange, programme designers hoped that these envisaged trades in carbon would encourage the exchange of other knowledge and values, including the repercussions of everyday activities and dependencies between the city and its hinterlands.

The City's net zero emissions imaginary predicted carbon markets as an inevitable and preferred response to climate change, and carbon offsets as an opportunity for local leadership and economic development within the city and its hinterlands. Carbon trading was designed to elicit support for the net zero emissions city by responding to local and global threats and creating numerous 'goods' – counteracting urban emissions and generating investment returns in the city, producing electricity, generating income and wider regional benefits, and strengthening relationships between urban and regional communities.

Stabilising net zero emissions: carbon offsets to renewable energy

The City of Melbourne updated its ZNE2020 strategy in 2008 and again in 2014. The updates continued to assert the net zero emissions city objective, even as emissions within the municipality climbed. By 2006, estimates of community-wide emissions had grown to 6.4 million tonnes and were anticipated to increase to almost eight million tonnes by 2020. The 2008 update set a goal to reduce these by around 1.5 million tonnes by 2020, with the remainder to be offset (City of Melbourne 2008, 22). However, in response to anticipated shifts in national climate policy, the City abandoned its proposal for a municipal carbon trading scheme. Instead, national commitments to establish emissions trading and renewable energy targets were expected to significantly reduce urban emissions.

The national government was seen by many as the appropriate level to deliver economically optimal strategies to reduce emissions with limited duplication and overlap (City of Melbourne 2014, 2). These expectations were reinforced by economic analyses from the UK and Australia that underlined the opportunities and benefits of proactive and early measures to reduce emissions and the costs of delayed action (Stern 2007; Garnaut 2008). The update considered it 'highly likely' that new market conditions created by these policies would result in 'investment to support the most economically viable, locally sourced, renewable and low-carbon technologies within the boundaries of the City of Melbourne' (2014, 2). Expectation of action by others reshaped the City of Melbourne's prospects for lowering emissions within municipal boundaries.

In terms of carbon accounting and offsetting, national methods and standards were further developed to generate trust in carbon markets, which were aligned to the Kyoto Protocol and the emergence of voluntary carbon trading. The National Carbon Offset Standard (NCOS) established an orthodox approach to carbon accounting for organisations, including local governments, and criteria for purchasing offsets generated in Australia and overseas that met basic criteria for accountability (Department of Environment and Energy 2010). This standard, alongside other developments in carbon accounting and reporting such as the *National Greenhouse and Energy Reporting Act 2011*, came to define the technical benchmarks against which the City of Melbourne and most other Australian local government organisations assert their carbon neutral status (City of Melbourne 2016).

Against these developments, the City of Melbourne abandoned its pilot programme to sequester carbon in regional Victoria in favour of purchasing carbon offsets via voluntary markets. In the 2011–2012 financial year, it was certified as a carbon-neutral organisation under NCOS. To meet this certification, the city administration purchased almost 10,000 MWh of 'Greenpower' under the national renewable energy target (negating over 13,000 tonnes CO_2^{-e}) and over 44,000 tonnes of carbon offsets. These were mainly carbon credits from capturing landfill gas under the national government's *Carbon Credits (Carbon Farming Initiative) Act 2011* and also included 200 tonnes CO_2^{-e} of offsets from a small-scale hydropower project in China accredited under the international Verified Carbon Standard programme. Alongside persistent efforts to reduce its corporate emissions, the City continued to offset remaining emissions each financial year from a range of projects in other countries, including Indonesia, Mali, Turkey, Zimbabwe, Uganda, and China (City of Melbourne 2016; Department of Environment and Energy 2019).

While these offsets related to corporate rather than community-wide emissions, they show that the City of Melbourne's preferences to trade urban carbon to a known and specific hinterland and organised with trusted partners were supplanted by an even more commoditised approach. The City's ability to carry out this change relied on a global system of standards, verification, regulation, and oversight to generate trust between market participants who are largely unknown to each other. The City's participation in these global markets displaced urban carbon farther away and mediated these trades through impersonal relationships, making the environmental, social, and economic impacts and implications of these trades much harder to discern.

In 2014, The City of Melbourne again updated ZNE2020 during a period of significant upheaval in the Australian landscape of climate and energy policy. National measures to put a price on carbon pollution and invigorate renewable energy investments did not provide the policy stability that many were hoping for. Australia's nascent emissions trading scheme was repealed in July 2014, followed a year later by cuts to the national renewable energy target (Kuch 2015; Talberg and Workman 2016). The 2014 update to ZNE2020 acknowledged that the first update 'was written with the assumption that Australia would put a price on carbon and international policy that would … drive significant emissions reductions' (City of Melbourne 2014, 3). The update stressed that action by the City of Melbourne alone would not achieve the target for zero net emissions, and 'the actions outlined in this strategy must be accompanied by fundamental changes to our energy supply which is subject to Australian and Victorian Government policy' (City of Melbourne 2014, 2). This statement stresses the extent to which the City's actions have been constrained by the decisions and actions of state and national governments.

Although the actions of those outside the municipality failed to deliver expected emissions reductions within the city, the 2014 update reasserted the commitment to net zero emissions. Looking beyond national politics, it affirmed the 'inevitability of the low carbon economy' and emerging opportunities to manage carbon within and beyond municipal boundaries (City of Melbourne 2008, 10). In this regard, the 2014 update demonstrated a range of approaches to lowering the Council's own emissions, and across key sectors in the municipality, including environmentally sustainable building and retrofitting practices for its offices and community buildings and support for other building owners to do the same.

Carbon offsets remained a central part of the 2014 update, albeit with some reservations. The 2014 update estimated that offsetting community-wide emissions in 2020 would cost around AUD\$30 million (City of Melbourne 2014, 3). This would be a recurring cost and would vary depending on the success of other measures such as renewable energy uptake, cutting emissions from transport and waste, and the cost of carbon offsets into the future. Offsetting urban carbon at such a scale would also defer more fundamental changes to socio-technical systems, lifestyles, and behaviours in favour of crediting actions taken elsewhere. This awareness served to dilute the economic imperatives for action defined in earlier iterations of the strategy. Despite their continued place in the City's climate strategy, carbon offsets appeared to be in tension with the net zero emissions city imaginary.

In contrast, policy actions concerned with decarbonising the energy supply were strengthened in the 2014 update. The City of Melbourne scaled up its renewable energy target for council operations to 25 per cent by 2018. To meet this target, it formed a consortium with other inner-city organisations to purchase renewable energy directly from a new wind farm in regional Victoria. Called the Melbourne Renewable Energy Project (MREP), a dozen large organisations negotiated to contract power directly from the proposed 80 MW Crowlands wind farm, enabling the proponent to secure finance and start construction. The City achieved 100 per cent renewable energy for its corporate operations when the wind farm came on line in January 2019 (Vorrath 2019). Policy-makers within the City of Melbourne described how the MREP circumvented the local government's lack of authority over energy policy by taking a demand-driven approach, whereby large energy consumers influence the composition of energy supply and mediate their energy demands through energy efficiency measures (Fieldnotes, March 2017). They recognised that 'local government has no policy levers to influence the composition of the grid, so they started looking at the demand side for options to bring new renewable energy generation online' (Research interview, December 2018).

The MREP project was developed at a time when the dominant industry generators and retailers were reluctant to contract new renewable energy because of uncertainties over the future direction of national energy policies, especially in regard to regulating GHG emissions. It enabled 'cities, corporations and institutions to take an active role in securing renewable electricity supply and taking action on climate change' (City of Melbourne 2017, 6). It also helped these customers 'mitigate the risk of increased energy costs in a volatile market with long-term price certainty' (City of Melbourne 2017, 6). The project positioned the City of Melbourne in a brokerage role within this uncertain landscape, facilitating complex arrangements between large organisations with widely varying needs (Research interview, December 2018).

The MREP also re-established the envisaged relationship between the city and its hinterlands in managing urban emissions, this time through renewable energy rather than carbon offsets. Tenders for the electricity contract were extended to sites in regional Victoria but not elsewhere. The project was presented as a way to share the wealth and prosperity of the city through regional development while at the same time producing 'home grown' power and 'local' energy (City of Melbourne 2017, 9). This reflects a persistent desire to recognise dependencies between the city and its hinterlands. Even so, the City and its partners in the project were unable to direct specific preferences for the location of the wind farm. The Crowlands site was selected from a range of proposals through a tender process so that all the suppliers were treated fairly (Research interview, December 2018).

The development of technical standards and global markets in carbon trading reconfigured the way the City of Melbourne achieved carbon neutrality for its corporate emissions. However, the accounting standards that established carbon as a tradeable commodity also limited possibilities for more relational displacements of carbon. The particular relations underpinning municipal carbon trading envisaged in the 2003 strategy – with rural municipalities and eucalyptus trees – were altered as carbon itself was detached from these tight relational bonds. Arrangements for renewable energy appear somewhat different in the MREP project, which brought together certain actors in urban and regional settings, albeit with these relations mediated through complex energy market contracts and large-scale electricity infrastructure.

Re-imagining the net zero emissions city

While the MREP enabled the City of Melbourne to realise zero net emissions for its own operations, achieving net zero emissions by 2020 across the entire municipality seems increasingly unlikely. In 2017, emissions across the municipality were over 4.6 million tonnes CO_2^{-e} (City of Melbourne 2018, 12). In November 2018, the City released its new *Climate Change Mitigation Strategy to 2050: Melbourne together for 1.5°C* (City of Melbourne 2018). The new strategy reasserts the City's commitment to community-wide net zero emissions but shifts the timeframes to act and the mechanisms to achieve the goal while also opening up possibilities for new approaches to municipal carbon accounting.

The period since the 2014 update of ZNE2020 saw significant changes in climate governance across regional, national, and international registers. At the national level, the turbulence of Australia's climate and energy policy landscape continued with unstable energy policies and unsatisfactory measures to reduce emissions in line with the Paris Agreement (OECD 2019). Against this, commitments and activities at sub-national and local levels gained momentum. Notably, the Victorian Government introduced state-based targets for renewable energy and a goal for statewide net zero by 2050 through the *Climate Change Act 2017*. At the international level, C40 Cities developed a platform for its member cities to reduce emissions in line with the Paris Agreement's 1.5°C target. Called Deadline 2020, the programme couples top-down modelling of global carbon budgets with assessments of cities' per capita emissions and GDP to define community-wide emissions reduction trajectories and identify priority actions (C40 Cities and Arup 2016, 10). The programme requires all its member cities to develop climate action plans before 2020 that are guided by these targets but configured around their particular circumstances.

The City of Melbourne was one of eight cities to pilot the Deadline 2020 programme as a basis for its new climate strategy (C40 Cities 2018). Deadline 2020 coordinates narratives of action in relation to the common target of 1.5°C and defines 'science-based' emissions reduction pathways of C40 member cities based on a 'contract and converge' model of equal per capita of emissions by 2030 (C40 Cities and Arup 2016, 7). In one sense, the Deadline 2020 programme abstracts cities from their settings to situate them against other cities around the world. This allows local governments to benchmark against others and raise the bar on their own climate governance, justifying and promoting stronger ambition. The actions required to meet these targets are expected to vary from one city to the next, but the overarching frame is one of ecological modernisation, where urban socio-technical infrastructures are transformed and resource flows secured in response to the risks of climate change.

The City's *Climate Change Mitigation Strategy to 2050* extends and adjusts the temporal boundaries of the net zero emissions city imaginary. Notably, aligning the strategy to the Paris Agreement and state government targets lengthens the timeframe to achieve net zero emissions appreciably from 2020 to 2050. The City was also the first local government in Victoria to pledge emissions reductions under the state government's *Climate Change Act 2017*. This alignment seems especially crucial given its limited policy levers to change wider systems of energy, transport, buildings, and waste. The emissions reduction targets 'can only be achieved through collaborative action with all three levels of government' (City of Melbourne 2018, 19).

Just as the City of Melbourne's new climate strategy reconfigures temporal boundaries, it also recasts the conception of spatial boundaries around the net zero emissions city across larger and smaller scales. The strategy opens up new possibilities for the City's municipal carbon accounting practices on several fronts. C40 Cities worked through the City of Melbourne to reconcile the municipal carbon accounts of all 32 of Melbourne's metropolitan councils to establish a community-wide emissions inventory for Greater Melbourne area (City of Melbourne 2018, 13, 61). The new strategy also mentions alternative consumption-based approaches to measuring emissions compared with territorial carbon accounts. Such an approach 'takes into account the upstream and downstream impacts of products and services that Melbourne consumes including imports and exports from the city' (City of Melbourne 2018, 62). The impacts of consumption are discussed throughout the strategy through the idea of the circular economy with respect to reducing waste and inefficiency and thus securing continued resource flows (City of Melbourne 2018, 17).

While these new carbon accounts stretch the conceptual boundaries around carbon in the city, many of the proposed actions establish boundaries of carbon neutrality around discrete spaces within the municipality. Certain urban renewal areas and buildings within the municipality are planned to showcase possibilities for net zero precincts and developments, guided, for example, by new NCOS standards for precincts (Commonwealth of Australia 2017) and the C40 Cities programmes on climate positive development (C40 Cities 2016). A proposed virtual power plant or solar garden (City of Melbourne 2018, 26) would reshape relations between certain residents and small businesses within the municipality by enabling electricity (and thus carbon) to flow in new directions.

Each of these adjustments serves to position the City of Melbourne as a key broker between multiple actors and across multiple levels of urban climate governance, from local residents, businesses, workers, and community groups to neighbouring municipalities, state and national government, and other cities around the world. A shift in language reflects this position. The political commitment to net zero emissions has moved from an 'end to the city's contribution to climate change' (City of Melbourne 2003, 3) to 'part of international efforts to avoid a 1.5°C increase in global temperatures … [aligned] with the science-based targets in the Paris Climate Agreement' (City of Melbourne 2018, 10).

Similar to previous strategies, the new strategy enlists economic modelling to assert that

> if we do not act decisively to reduce emissions as part of the global effort, the impacts of climate change and missed economic opportunities of

transitioning to a low carbon economy will cost the community $12.6 billion from 2020 to 2050.

(City of Melbourne 2018, 5)

Compared with this estimate, the net economic benefits of taking 'significant' or 'accelerated' action to reduce emissions are modelled at $3 billion and $5.6 billion respectively (City of Melbourne 2018, 5). The strategy rejects an option to achieve net zero emissions by 2020 by purchasing carbon offsets, based on high recurring cost (higher even than estimated in the 2014 update) and lack of local benefits. 'The estimated cost to the City of Melbourne of the Purchasing Offsets Scenario would be AU$240–480 million per year and would not address the underlying causes of emissions in the municipality' (City of Melbourne 2018, 5). In principle, the City accepts that offsets can be justified where they provide 'important environmental, social and economic benefits for reducing emissions in remote and regional Australia, and in many other countries' (City of Melbourne 2018, 22). However, in terms of relying on offsets to meet and sustain net zero emissions, 'achieving emissions reductions through purchasing offsets alone will not address the systemic causes of GHG emissions or achieve the full extent of benefits for Melbourne residents and businesses' (City of Melbourne 2018, 22).

Thus, the role of carbon offsets in the City of Melbourne's net zero emissions imaginary has been largely dismissed since its first iteration in 2003. The new strategy concedes that small amounts of residual emissions from transport and waste are likely, and anticipates that the viability of carbon offsets may improve over time as the international carbon market develops (City of Melbourne 2018, 20). Placing this in a broader context, the Deadline 2020 modelling for its member cities to meet their share of the 1.5°C target anticipates the need for negative emissions technologies such as large-scale bioenergy carbon capture and storage from 2050 and beyond (C40 Cities and Arup 2016, 102). In the shorter term, these assumptions appear immaterial to the development of C40 cities' climate action plans that stress the need for urban emissions to peak and decline within the next five to ten years (C40 Cities and Arup 2016, 31) and emphasise the use of offsets as a last resort after concerted efforts to reduce emissions (Ernst and Young 2018, 2). Even so, carbon offsets emerging in a somewhat new form of negative emissions technologies remain an enduring part of the imagined future of net zero emissions cities.

Conclusion

This chapter has explored ongoing tensions involved in urban climate governance through the City of Melbourne's successive imaginaries of the net zero emissions city. The City's efforts to assert and stabilise its political commitments to the net zero emissions target have involved pragmatic and continual adjustments to shifting circumstances.

The City's initial 2003 strategy for zero net emissions proposed a local market in carbon trading built on established relationships. Preferences to displace urban carbon to regional Victoria were eclipsed with the emergence of national and international standards for carbon neutrality and carbon trading. Yet these spatial preferences re-emerged through the MREP that explicitly set out to see new wind farms built in regional Victoria. This suggests a persistent desire by the City to balance the carbon accounts as locally as possible, if not by trading carbon between urban and regional municipalities then by procuring renewable energy within the State. The 2018 strategy maintains the emphasis on reducing local emissions as much as possible in preference to absolving those emissions by purchasing carbon offsets, even if it extends the time frame for net zero emissions to 2050. The strategy also reframes the net zero emissions city in relation to sub-national and international efforts towards global net zero emissions by 2050, and positions the City of Melbourne as a key actor and broker within the envisaged green transition of the greater Melbourne metropolis.

The City's ambition to lead on climate change has been tempered by its limited authority to effect changes to the extensive socio-technical systems through which urban GHG emissions are created. Opportunities to take actions outside the municipal boundaries through carbon offset strategies and renewable energy investments have been balanced against preferences to reduce emissions as locally as possible. The City's narratives of green urban transition have been pitched inwards towards urban businesses and residents but also outwards to other municipalities, levels of government, investors, city networks, and so on.

Through successive iterations of the City of Melbourne's climate strategy, targets for the net zero emissions city have largely been framed in techno-economic terms. National and international standards of carbon accounting and offsetting have been used to determine and legitimise carbon neutral certification. Science-based targets have apportioned global carbon budgets to discrete cities and entities. Cost-benefit analyses have framed and guided policy decisions. However, shifts in the City's preferences for policies and technologies to meet its target outlined above suggest that policy design and implementation are not only technical, but involve ongoing negotiation over the legitimacy and viability of local government involvement in urban climate governance.

These tensions articulate complex and dynamic relationships between the city and its hinterlands and varied interpretations of how far these relationships extend. In this sense, practices of accounting for and allocating carbon towards an ideal of a net zero emissions city are not only technical but also assign responsibilities, establish relationships, and reconfigure space in particular ways. The conceptual boundaries of the net zero emissions city are partial, permeable, and contestable. Highlighting these conceptual boundaries more explicitly rather than burying assumptions in technical and administrative processes may open up possibilities for more nuanced and locally situated economies to emerge around carbon – whether through trees, wind turbines, or other means – that acknowledge and add layers to these strands of connectivity.

References

Acuto, M. (2016) Give cities a seat at the top table. *Nature*, 537, 611–613.

Acuto, M. and Rayner, S. (2016) City networks: breaking gridlocks or forging (new) lock-ins? *International Affairs*, 92, 1147–1166.

Agarwala, M. (2015) Push to decarbonize cities after Paris talks (Correspondence). *Nature*, 528, 193.

Betsill, M.M. and Bulkeley, H. (2006) Cities and the multilevel governance of global climate change. *Global Governance*, 12, 141–159.

Blok, A. (2013) Urban green assemblages: an ANT view on sustainable city building projects. *Science and Technology Studies*, 26, 5–24.

Broto, V.C. (2017) Urban governance and the politics of climate change. *World Development*, 93, 1–15.

Bulkeley, H., Broto, V.C., and Maassen, A. (2014) Low-carbon transitions and the reconfiguration of urban infrastructure. *Urban Studies*, 51, 1471–1486.

Bumpus, A.G. and Liverman, D.M. (2008) Accumulation by decarbonization and the governance of carbon offsets. *Economic Geography*, 84, 127–155.

C40 Cities (2016) Good practice guide: climate positive development. Available at: www.c40.org/other/climate-positive-development-program.

C40 Cities (2018) C40 climate action planning resource centre. Available at: https://resourcecentre.c40.org/climate-action-planning-framework-home.

C40 Cities and Arup (2016) Deadline 2020: how cities will get the job done. Available at: www.c40.org/other/deadline_2020.

Callon, M. (2009) Civilizing markets: carbon trading between in vitro and in vivo experiments. *Accounting, Organizations and Society*, 34, 535–548.

Carbon Neutral Cities Alliance. Available at: https://carbonneutralcities.org.

City of Melbourne (2003) Zero net emissions by 2020, a roadmap to a climate neutral city. Available at: www.melbourne.vic.gov.au/SiteCollectionDocuments/zero-net-emissions-2002.pdf.

City of Melbourne (2008) Zero net emissions by 2020 update 2008. Available at: ttp://www.melbourne.vic.gov.au/SiteCollectionDocuments/zero-net-emissions-2008.pdf.

City of Melbourne (2014) Zero net emissions by 2020 update 2014. Available at: www.melbourne.vic.gov.au/SiteCollectionDocuments/zero-net-emissions-update-2014.pdf.

City of Melbourne (2016) Emissions reduction plan for our operations 2016–2021. Available at: www.melbourne.vic.gov.au/SiteCollectionDocuments/emissions-reduction-plan.pdf.

City of Melbourne (2017) Melbourne renewable energy project: a new generation of energy. Available at: www.melbourne.vic.gov.au/business/sustainable-business/mrep/Pages/melbourne-renewable-energy-project.aspx.

City of Melbourne (2018) Climate change mitigation strategy to 2050: Melbourne together for 1.5°C. Available at: www.melbourne.vic.gov.au/about-council/vision-goals/eco-city/Pages/climate-change-mitigation-strategy.aspx.

City of Melbourne (2019) Melbourne facts and figures. Available at: www.melbourne.vic.gov.au/about-melbourne/melbourne-profile/Pages/facts-about-melbourne.aspx.

Commonwealth of Australia (2017) National carbon offset standard for precincts. Available at: www.environment.gov.au/system/files/resources/91aadf60-1454-4cde-81dd-587df7cdadd7/files/ncos-precincts.pdf.

Davis, S.J. and Caldeira, K. (2010) Consumption-based accounting of CO_2 emissions. *Proceedings of the National Academy of Sciences USA*, 107, 5687–5692.

Department of Environment and Energy (2010) *National Carbon Offset Standard*. Available at: www.environment.gov.au/climate-change/government/carbon-neutral/ncos.

Department of Environment and Energy (2019) *National Carbon Offset Standard Certified Organisations*. Available at: www.environment.gov.au/climate-change/government/carbon-neutral/certified-businesses/city-melbourne.

Dodman, D. (2009) Blaming cities for climate change? An analysis of urban greenhouse gas emissions inventories. *Environment and Urbanization*, 21, 185–201.

Ernst and Young (2018) City of Melbourne climate action planning technical assistance: synthesis report. Available at: https://participate.melbourne.vic.gov.au/climatechange.

Garnaut, R. (2008) *The Garnaut Climate Change Review: Final Report*. Port Melbourne, Vic.: Cambridge University Press.

Geden, O. and Beck, S. (2014) Renegotiating the global climate stabilization target. *Nature Climate Change*, 4, 747–748.

Gleeson, B. (2014) *The Urban Condition*. London and New York: Routledge.

Greenhouse Gas Protocol (2016) Global protocol for community-scale greenhouse gas emission inventories: executive summary. Available at: www.ghgprotocol.org/sites/default/files/ghgp/standards_supporting/GPC_Executive_Summary_1.pdf.

Hodson, M. and Marvin, S. (2017) Intensifying or transforming sustainable cities? Fragmented logics of urban environmentalism. *Local Environment*, 22, 8–22.

Jasanoff, S. (2015) Future imperfect: science, technology, and the imaginations of modernity. *Dreamscapes of Modernity: Sociotechnical Imaginaries and the Fabrication of Power*, 1–47.

Jasanoff, S. and Kim, S.H. (2009) Containing the atom: sociotechnical imaginaries and nuclear power in the United States and South Korea. *Minerva*, 47, 119.

Jasanoff, S. and Kim, S.H. (2015) *Dreamscapes of Modernity: Sociotechnical Imaginaries and the Fabrication of Power*. Chicago, IL: University of Chicago.

Jasanoff, S., Wynne, B., Buttel, F., *et al.* (1998) Science and decisionmaking. In S. Rayner and E.L. Malone (eds), *Human Choice and Climate Change*. Columbus: Battelle Press.

Jordan, A.J., Huitema, D., Hildén, M., *et al.* (2015) Emergence of polycentric climate governance and its future prospects. *Nature Climate Change*, 5, 977–982.

Kenis, A. and Lievens, M. (2017) Imagining the carbon neutral city: the (post)politics of time and space. *Environment and Planning A*, 49(8), 1762–1778.

Knutti, R., Rogelj, J., Sedlacek, J., *et al.* (2016) A scientific critique of the two-degree climate change target. *Nature Geoscience*, 9, 13–20.

Kuch, D. (2015) *The Rise and Fall of Carbon Emissions Trading*. New York: Springer.

Liu, Z., Feng, K., Hubacek, K., *et al.* (2015) Four system boundaries for carbon accounts. *Ecological Modelling*, 318, 118–125.

MacKenzie, D. (2009) Making things the same: gases, emission rights and the politics of carbon markets. *Accounting, Organizations and Society*, 34, 440–455.

Marvin, S. and Hodson, M. (2010) *World Cities and Climate Change*. London: McGraw-Hill Education (UK).

Miller, C.A. and Edwards, P.N. (2001) *Changing the Atmosphere: Expert Knowledge and Environmental Governance*. Cambridge, MA: MIT Press.

OECD (2019) Environmental performance review of Australia. Available at: www.oecd.org/australia/australia-needs-to-intensify-efforts-to-meet-its-2030-emissions-goal.htm.

Satterthwaite, D. (2008) Cities' contribution to global warming: notes on the allocation of greenhouse gas emissions. *Environment and Urbanization*, 20, 539–549.

Schleussner, C.-F., Rogelj, J., Schaeffer, M. *et al.* (2016) Science and policy characteristics of the Paris agreement temperature goal. *Nature Climate Change*, 6, 827–835.

Stern, N. (2007) *The Economics of Climate Change: The Stern Review.* Cambridge, UK: Cambridge University Press.

Talberg, A. and Workman, A. (2016) Timeline: Australian climate and clean air policy interventions 2013–16. Australian-German Climate & Energy College. Available at: www.climate-energy-college.net/files/site1/docs/6561/WorkingPaper_1_14June2016.pdf.

Thompson, M. and Rayner, S. (1998) Risk and governance part I: the discourses of climate change. *Government and Opposition*, 33, 139–166.

UNEP (2018) Emissions gap report. Available at: www.unenvironment.org/resources/emissions-gap-report-2018.

UNFCCC (2015) Paris Agreement. United Nations treaty collection. Available at: https://treaties.un.org/doc/Treaties/2016/02/2016021506–03PM/Ch_XXVII-7-d.pdf.

Vorrath, S. (2019) City of Melbourne hits 100% renewables as 80MW wind farm comes online. *One Step Off the Grid.* Available at: https://onestepoffthegrid.com.au/city-of-melbourne-hits-100-renewables-as-crowlands-wind-farm-comes-online/.

13 Governing the transnational

Exploring the governance tools of 100 Resilient Cities

Anne Bach Nielsen

Introduction

In the southern part of Denmark lies a small picturesque city named Vejle. The city is located at the head of a fjord, where two rivers and their valleys converge. In recent years, Vejle has begun to witness the implications of climate change, urbanisation trends and new immigration flows, and consequently is now realising that new policies are necessary if the city is to address and adapt to these present and future challenges. The City of Vejle produced and released a political strategy to create citywide resilience to these challenges in the spring of 2016.

More than 9000 km away from Vejle, Mexico City, the capital of Mexico, also struggles with the conjuncture of a changing world. Mexico City is fundamentally different from Vejle in terms of geography and economic and political characteristics. Yet, it, too, has developed a resilience strategy in 2016 whose goal is a better future for its urban population. What do the two cities have in common that prompted them to create local resilience strategies? The answer is quite straightforward: they are members of the same network organisation – 100 Resilient Cities (100RC).

Neither Vejle nor Mexico City is unique in this regard. Across the world and on every continent, cities large and small are joining forces in so-called transnational municipal networks (TMNs) (Betsill and Bulkeley 2006; Keiner and Kim 2007; Bulkeley and Newell 2015; Acuto and Rayner 2016). Existing literature on TMNs offers conceptual work on the function and impact of TMNs (Betsill and Bulkeley 2006; Andonova *et al.* 2009; Andonova and Tuta 2014; Nielsen 2016; Busch 2016). They offer an understanding of how TMNs operate across scales, sectors, and actors, and how TMNs transcend our conventional understanding of what is global and local, public and private. However, despite their initial desire to understand the embedded power structures of TMN governance (Bouteligier 2012; Bulkeley and Newell 2015), scholars continue to offer a limited grasp of the nature of the governance tools provided by these networks.

In this chapter, I explore the type of governance tools and compliance mechanisms provided by TMNs through an analytical framework that I apply to the case of 100 Resilient Cities. Governance tools are here understood as interventions made by the 100RC to achieve central network rules and network goals. By applying insights from the vast body of literature on governance thinking, I suggest a framework that opens up the nuances of soft tools, which are usually used to describe the governance measures available to TMNs. Much of the existing literature on TMNs depicts TMN governance as voluntary, horizontal, and polycentric and without hard measures to conform members in certain directions. I argue, however, that TMNs can be powerful actors with capabilities to enforce membership rules among city members.

The contribution of the chapter is threefold. First, the chapter reviews the literature on TMNs through the lens of how previous studies have understood governance through multi-level frameworks and network theory. Second, the chapter proposes a framework for analysing TMN governance tools. I develop this framework based on the vast body of literature discussing soft and hard measures of governance. This framework aims to identify where and by whom decisions are made in a city-to-network relationship, and thus potentially increases our understanding of the diverse landscape of TMNs and the differences in their governance interventions. Third, the chapter provides empirical insight into the 100 Resilient Cities (100RC), which is a relatively new TMN that we still do not know much about. The 100RC is a global network sponsored by The Rockefeller Foundation, a private philanthropy, which has had tremendous success initiating similar resilience policies across its member cities. 100RC simultaneously illustrates the applicability of the analytical framework.

The chapter proceeds as follows: in the following section I examine the existing literature on how transnational municipal networks are approached in frameworks of multi-level governance. Building on this body of literature, I develop the analytical framework based on previous thinking on governance tools. I then turn to the methodology before I analyse and discuss the governance tools provided by the 100RC. Finally, I offer conclusions based on the analysis and raise a set of critical questions which are important for our further understanding of TMNs.

Governing the transnational space: learnings from the academic literature

TMNs are formalised networks of local authorities, in most cases municipalities, which cooperate and operate beyond national borders in network-like configurations (Bulkeley *et al.* 2003). Even though city networking is not a new phenomenon, 60 per cent of the active networks are less than 30 years old (Acuto and Rayner 2016, 1156). The significant recent increase in networks may be explained by the fact that national, international, and transnational actors are increasingly recognising the local level as the source of important solution-providers for some of the world's greatest challenges (Betsill and Bulkeley 2006; Rayner 2010; Barber 2013; Hoff and Gausset 2015). Municipal governments have both the knowledge and authority to influence critical areas such as land use planning, waste management, energy supply and consumption, housing and public transportation. In order to develop knowledge about the heterogeneous landscapes of TMNs, researchers have explored the overall characteristics of TMNs (Bulkeley *et al.* 2003; Andonova *et al.* 2009), the internal dynamics (Keiner and Kim 2007), and the impacts of city networks on urban governance initiatives and structures (Kern and Bulkeley 2009; Busch 2016; Rashidi and Patt 2018).

Much of this literature builds on ideas of multi-level governance and changes in the global distribution of power (Betsill and Bulkeley 2006; Bulkeley and Newell 2015; Busch 2016). Betsill and Bulkeley (2006, 145) argue that dominating approaches to understanding global environment governance take their departure from state-centric assumptions, which do not take into account other emerging modes of governance. Instead, they suggest a multi-level approach originally developed in the context of the European Union (Hooghe and Marks 1996, 2001), which recognises the value of including sub-state and non-state actors in environmental governance. Multi-level governance gives voice to complex and polycentric allocation of authority across domestic and international levels. It does so by acknowledging the existence of another mode of governance known as Type II governance, which escapes the conventional divide between national and international levels and emphasises that decisions affecting world politics can be made outside the sphere of nation-states and international regimes (Gupta *et al.* 2015). In this view, transnational actors can organise beyond the levels of territorial jurisdictions in order to act collectively and negotiate across boundaries (Hooghe and Marks 2001). Here, political authority is diffused and horizontal structures of cooperation may coexist beside one another and along more conventional modes of vertical governance structures (Bulkeley and Newell 2015). Type II governance is thus not bound to a particular scale (Betsill and Bulkeley 2006) or to a particular form of actor (Bulkeley and Newell 2015).

This view of global governance has broadened our understanding of the realm that TMNs work in. TMNs appear in public, hybrid, and private forms of governance (Betsill and Bulkeley 2006; Bulkeley and Newell 2015; Acuto and Rayner 2016). They are local and global, and their actions are easier to capture by looking at relationships and interaction than by looking at presumed levels of authority (Acuto 2013). Building on these insights, scholars describe membership of TMNs as voluntary constellations where members are free to leave a network at any time (Kern and Bulkeley 2009; Andonova *et al.* 2009). Because these networks are polycentric and horizontal in structure, they are characterised by a mode of internal self-governance (Betsill and Bulkeley 2006). As a consequence, and despite explicit ambitions of creating environmental action, the measures available to TMN are soft in nature and often left to the members to adopt and implement (Bulkeley and Newell 2015).

Despite the existing body of literature on TMNs, there are only a few studies that assess the nature of the toolkits deployed by networks in order to steer members towards network goals. How do TMNs share information, set rules, build capacity, and implement policies? The literature on TMNs gives us a broad sense of the governance practices that TMN members carry out, but we know very little about the mechanisms put in place to enforce compliance among members. To fill some of this gap, I suggest an analytical framework for analysing the governance tools provided by TMNs.

Governance tools

> In the absence of traditional governance toolkits of nation-states and international institutions, transnational governance arrangements have devised a range of means through which they seek to guide and direct their constituents towards particular goals.
>
> (Bulkeley and Newell 2015, 72)

Many TMNs are characterised by what Lee and Jung (2018, 98) refer to as institutional-led cooperation, which often requires a more formal organisation to coordinate dialogue, decision-making, and other network activities. These formal institutions have various designs from TMN to TMN but may include a physical secretariat, employees, a set of official membership rules as well as an organisation strategy and a list of network goals. Most commonly, researchers have analysed these TMN organisations within the frame of soft tools. Placed within the realm of Type II and often characterised by non-hierarchical structures, self-governance, and voluntarism, traditional hard tools seem to have a limited fit in the realm of TMN governance. But what are those soft measures and how do they guide members toward network goals? Despite the seemingly voluntary and polycentric sense of authority, TMNs are not powerless actors without measures to move members in a certain direction. Even though centralised TMN measures are considered soft, they are often designed, framed, and implemented by the network's central organisation. By applying insights from the vast body of literature on governance tools, I suggest an analytical framework that opens up the nuances of the soft tool framework.

It is a common approach to distinguish between different types of policy tools when analysing the characteristics of the governance structure in question. Traditionally, the distinction between hard and soft tools has been widely applied in the literature (Abbott and Snidal 2000; Shaffer and Pollak 2009; Blomqvist 2016). This distinction usually refers to a distinction between binding and non-binding obligations in legal terms (Abbott and Snidal 2000). Hard tools are commonly associated with legislation, and soft tools are associated with non-binding arrangements such as recommendations, guidelines, information, and goal-setting (Kirton and Trebilcock 2017). These kinds of tools are considered soft because compliance cannot be enforced through hard sanctions such as trade embargos or court verdicts. Moreover, hard tools are often more fixed in terms of clear and precise rules of conduct while soft tools are more ambiguous and open to interpretation at the implementation level (Abbott *et al.* 2000). A third central dimension is the distinction between

Table 13.1 Characteristics of hard versus soft governance tools

Hard tools	Soft tools
Legal obligation	No legal obligation
Fixed	Flexible
Authoritative	Deliberative

Source: developed on the basis of Blomqvist (2016, 269).

authoritative and deliberative steering logics (Blomqvist 2016). An authoritative instrument represents a hierarchical power structure with a top-down steering logic; soft instruments reflect more bottom-up and participatory approaches to the development of the tool. The three classical defining characteristics of hard and soft governance are summarised in Table 13.1.

Even though hard and soft tools are often presented as binary categories, hard and soft measures are more likely a question of scale. Few tools are strictly hard or strictly soft: they reflect a degree of hardness/softness or a combination of hardness and softness along the dimensions presented in the table. It is clear, however, that TMNs do not deploy toolsets that are hard in the most conventional and strict sense: they do not have the authority to design legal instruments or make use of hard sanctions. However, by opening up the softness of TMN governance, we get a fuller picture of how TMNs interact with society and how their members respond to the activities offered by the network.

I suggest an analytical framework building on the categories introduced in Table 13.1. First, the extent to which TMN tools are voluntary in nature should be considered. Networks vary considerably in this regard, and many resemble what Green (2017) refers to as pseudo-clubs: governance initiatives where the membership is fluid, almost free, and where benefits are limited. This type of TMN rarely imposes any kind of obligations on their members or sanction non-compliance with network obligations. In other cases, networks can exert power over their members and sanction misconduct of central obligations (Bulkeley and Newell 2015; Rashidi and Patt 2018). Instead of applying a criterion of legal obligation, membership obligations and the use and non-use of sanction mechanisms may tell us something about the governance structures.

Second, the steering logic behind a governance tool might vary across both TMNs and across the tools provided internally by one TMN. Two important factors may be considered.

(1) Is the tool developed from a participatory approach where members are co-influencing, or even taking the initiative to develop a measure, or does the tool-development reassemble an elitist approach where few actors frame the content, implementation, and evaluation schemes?

(2) Which actors (with what interests) participate in the tool-development with what content?

In a study of the Asian Cities Climate Change Resilience Network (ACCCRN), Stephen Tyler and Marcus Moench (2012) analyse a conceptual framework put forward by the network organisation. The ACCCRN was, like 100RC, initiated by The Rockefeller Foundation, and is committed to kick-start climate adaptation planning and implementation in East Asian cities over a five-year period through partnerships of local, regional, and national organisations. The ACCCRN secretariat, together with the Institute for Social and Environmental Transition, developed a framework and methodology for planning for climate resilience in all the member cities. Cities were then expected to apply the conceptual framework in their local work with climate adaptation planning (Tyler and Moench 2012). Jens Hoff (2018) discusses this framework and describes it as a benign top-down approach to climate adaptation policies. However, despite acknowledging elements of civil society engagement as part of the conceptual framework (in the implementation phase), the approach still embeds a hierarchical steering logic where the developed framework is implemented top-down (Hoff 2018). Tyler and Moench (2012) demonstrate that by determining the criteria of both member participation and framework development and by providing certain content in central network tools, TMNs become powerful actors.

Third, the flexibility of the tool can be important for understanding the nature of the tool. Is the tool in question loosely defined and translated (and maybe even transformed) into the local context of the network member? Or is the tool 'fixed' and meant for direct implementation into the political context of the member city with guidelines on how the framework should be interpreted and implemented?

By applying these three dimensions – membership obligation, steering logic, and interpretive flexibility – we broaden our understanding of TMN goal compliance. The framework also raises critical questions about power dynamics, legitimacy, and accountability when the obligation patterns and the steering logic behind a tool are analysed more closely.

Methodology

Much of the existing literature on TMNs takes its departure from a European context and most of the literature is written about networks that specifically focus on climate change issues – so-called transnational municipal climate networks (TMCNs) (Bulkeley *et al.* 2003; Bulkeley and Kern 2006; Keiner and Kim 2007; Andonova *et al.* 2009; Busch 2016). The strong attention paid to climate change reflects the TMN landscape to some extent as most of the existing networks point to climate change as their core agenda. This is not the full empirical picture, however, as many TMNs direct their attention towards other purposes such as peace-building, poverty reduction, equality concerns, and cultural agendas (Acuto and Rayner 2016, 1154). I argue along the lines of Michelle Acuto and Steve Rayner (2016) and stress the importance of a more systematic and full coverage of TMNs that reaches beyond geographical boundaries and topical areas.

My initial inquiry into the politics of TMNs began as an explorative search for a network that could provide a global and topic-wide insight to city-networking. I quickly chose the recently established network 100 Resilient Cities because it reflected those qualities and allowed me to follow how governance structures were put in place from the beginning. In addition, the 100RC network serves as an extreme case for the questions asked in this paper (Flyvbjerg 2001) because it has some of the strictest conditions for membership among the existing TMNs (more on this in the following section). In terms of material, my analysis of 100RC builds on two main sources.

First, the analysis builds on 30 formal interviews and many informal conversations conducted in the three cities (Vejle, Denmark; Chennai, India; and Porto Alegre, Brazil) as well as with 100RC regional offices. For this particular analysis, I draw on interviews with the Chief Resilience Officer (CRO) placed in each 100RC member city and from my conversations with representatives from the 100RC offices. The interviewees' names are anonymised; however, I indicate their position and organisation. Second, I quickly discovered the substantial amount of political documents produced on the topic and took the opportunity to look across political contexts by comparing the large amount of political strategies. More precisely, I included 21 resilient strategies produced by city governments as a direct result of their membership of 100RC. In addition, I have analysed official 100RC documents, including 100RC official goals and guidelines. All of the documents and interviews are in English, Portuguese, or Danish, which allowed me to analyse them with limited help from an interpreter. I made a thematic analysis of both interviews and documents (see Bowen 2009; O'Leary 2014), where I coded the interviewees' responses according to emerging themes and categories. This approach supports a more open process that is relatively free of pre-determined categories, while narrowing down the many thousand pages of documents to the research questions I was interested in.

Exploring the 100RC toolbox

The analysis of the 100RC governance toolbox is structured in three overall sections. I first introduce the 100RC network and its goals and key activities. Having established these, I go a step deeper into the empirics and then analyse the toolkit deployed by the 100RC to secure city compliance with the network functions.

Introducing 100 Resilient Cities

100 Resilience Cities was officially launched in 2013 when the first group of 32 cities were accepted to the network. In 2014, an additional 35 cities were accepted, and a third group of 33 cities was added in May 2016 for a total number of 100 cities (100RC 2018). According to the 100RC, the core mission of the network is to build a global movement that acts as a catalyst for resilience

in cities, where resilience is broadly understood 'as the capacity of individuals, communities, institutions, businesses, and systems within a city to survive, adapt, and grow, no matter what kinds of chronic stresses and acute shocks they experience' (100RC 2018).

100RC has members across the world and works with both small and larger cities, which reflects a marked degree of global inclusiveness. Cities are accepted based on an application called the City Resilience Challenge, which is judged by 100RC and a panel of expert judges comprised of 100RC partners and stakeholders. Membership is competitive and the application process is also costly in terms of both time and money. The assessment panel looks at a range of criteria and assesses each application city's potential as a 'most likely' fit with 100RC. These criteria include the mayor's commitment, the city's history of dealing with 'shocks' or stress factors, its history of building partnerships, and its capacity to work across sectors in partnerships (100RC 2018).

The 100RC is first and foremost a capacity-building network and does not offer direct lobbying or advocacy services to its members. In an interview with a director from one of 100RC's regional offices, the interviewee distinguished between a political network and an implementation network to differentiate between the many kinds of network activities found across TMNs. The interviewee clearly places 100RC in the implementation category:

> It is not a political network; it is an implementation network to a very large extent. And what do I mean by that? We try to help the cities change in the way they operate. Make them better equipped to deal with the challenges they identify. Other networks do advocacy for example. We will advocate for the resilience agenda through our CROs, but we will not advocate directly that cities should have more power in Brussels, for example. Our cities can use our work to do that type of work, but they will use different arenas than the 100RC.
>
> (Interview, Regional Director, 100RC)

The director further defined the 100RC implementation activities through three central dimensions: (1) consultancy, (2) deliberation, and (3) problem-solving. Consultancy refers to a range of services and institutional infrastructure that is integrated into the member cities. Every city is enrolled in an intensive resilience programme when accepted into the network. Here, the 100RC offers consultancy in terms of actual consultants' expertise, management frameworks and processes. There is a strong incentive for cities to cooperate by disseminating their knowledge, sharing best practices, and showcasing solutions. However, in order to successfully carry out subsequent deliberation activities – and for the network to have an effect on city practices – cities need a clear understanding of the issues ahead and a common language in which to deliberate. This is where the consultancy dimension plays a crucial part for the 100RC network.

Second, the 100RC offers a deliberation dimension that enables cities to share best practices and exchange knowledge. This is a very common network function and found in most TMNs and in the academic literature on the subject. In the context of the 100RC, deliberation is a way for cities to learn and adopt from others (scaling), to spark reflection on ongoing projects, and to showcase their own work (branding). Third, 100RC is strongly tied to what it calls the 'platform of partners', which is a broad palette of partners from both the private and public sectors willing to engage with the individual member city in order to develop concrete solutions to the problems they identify from working with consultancy and deliberation dimensions.

This particular understanding of the relationship between different network activities is important for the further engagement with the specific 100RC tools. Cities are enrolled in a two- to three-year strategy programme shortly after they are accepted to the network. In order to secure the city's development in accordance with this programme, the 100RC offers a range of specific consultancy tools. These tools are mandatory for the cities to adopt and applied systematically across the 100 resilient cities across the world; they are used in this analysis to apply the theoretical framework developed earlier in the chapter.

The 100RC toolbox

The 100RC offers a large range of governance tools: some are obligatory to adopt and others are voluntary or limited to selected members. However, member cities are faced with three broad governance tools when they first enter the network, which are designed to create a common network language on resilience building and to guide cities towards resilience from the very beginning. The three tools are described in detail in the following sections.

The Resilience Framework and the Strategy Process

The 100RC network takes it departure from the concept of resilience and the fact that cities are increasingly under pressure from shocks and stresses stemming from urbanisation, globalisation, and environmental degradation. Based upon the definition of urban resilience, 100RC has developed a Resilience Framework together with the global consultancy firm, Arup, which was in charge of conducting an analysis of the most important dimensions for building a resilient city. This has resulted in a framework comprising four dimensions and 12 drivers (Figure 13.1). The four dimensions are health and well-being, economy and society, infrastructure and environment, and leadership and strategy. All dimensions are developed to guide member cities when they design their resilience strategies.

The Resilience Framework is based on a detailed analysis where different stakeholders are invited to provide input; however, the framework is still a 'stranger' to the city adopting it. The steering logic underpinning the Resilience Framework is largely similar to the approach of the aforementioned Asian Cities Climate Change Resilience Network (ACCCRN). The framework is developed based on the same benign top-down approach as seen in the ACCCRN, and

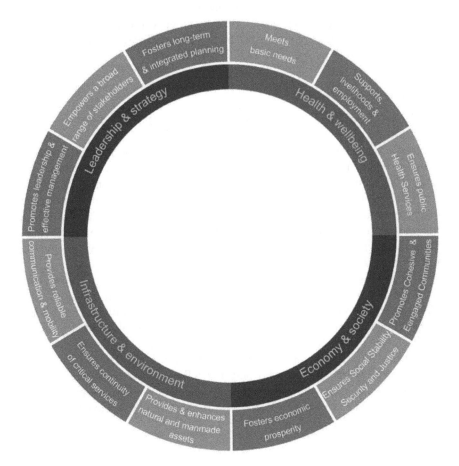

Figure 13.1 The City Resilience Framework.

Source: permission; Arup/Rockefeller Foundation (2014).

cities are expected to work with the framework as a guiding tool. However, as I discuss later, because the implementation and evaluation process of the framework is left to the member cities, the tool is thus extremely flexible and often largely transformed when it enters the local context.

The Strategy Process is also a predesigned process that the member city carries out over two phases. The first phase is kicked off with an agenda-setting workshop and is designed to set the agenda of the resilience work in the city. This includes data gathering, community and stakeholder engagement, and mapping out existing projects and potentials in a city. The product following phase one is a 'Preliminary Resilience Assessment', which is an official document portraying the current 'landscape' of risk and opportunities in the city. In the second phase, the city works actively with the strategy development by turning the assessment from phase one into tangible initiatives and projects. Phase two results – most importantly – in a Resilience Strategy, which is a defining political strategy co-signed by 100RC and the mayor.

The Chief Resilience Officer

The 100RC finances a Chief Resilience Officer (CRO), who is hired directly to manage the city's membership of the organisation. The CRO is placed in the member city and in charge of managing the process of the resilience process, which is a 6- to 12-month effort that results in the production and release of a city-specific resilience strategy. The CRO is fully funded by the 100RC and a city receives up to one million US dollars to finance the CRO and some of the activities around the strategy process. The CRO is supposed to function as a top-level advisor to the city's political authority and is described as 'the Mayor-1' by many of my interviewees. This position is a comprehensive attempt to institutionalise the network within the municipality.

The CRO is essential to the city's membership: she is in charge of managing the strategy process that leads to the development of a resilience strategy. The CRO is also responsible for making use of the tools made available through the network, including the City Resilience Framework and the platform of partners.

The 100RC engages with the member city substantially during the first years of membership, either by direct representation through the CRO, or through a number of strategy partners, who help the CRO with the management and execution of resilience goals. In the case of Vejle, for example, the strategy partners were a consultant from Arup and a consultant from Local Governments for Sustainability (ICLEI).

The Resilience Strategy

In most cases, the Resilience Strategy is a combination of an introductory risk assessment and an action plan on how the city plans to meet these risks. The Resilience Strategy is thus a culmination of the work done in both phases of the Strategy Process. My analysis of the existing resilience strategies reveals that the overall framing and format in each strategy are very much the same – everything from the broad narrative of a globalised, urbanised, and complex world under great pressure to more or less the same definition of resilience with a particular emphasis to treat resilience as an opportunity and as an essential good capacity to have in a city. Almost all strategies clearly refer to the Resilience Framework and describe or visualise how the framework dimensions and drivers are translated into the specific city.

The analysis also reveals, however, that the strategies are shaped by the heterogeneous political realities in each city. Almost all cities point to weather-related risk as a key challenge, which is directly linked to both climate change and disaster risks. Likewise, all of the cities point to urbanisation trends as one of the biggest challenges today and in the future. This concern is translated into more concrete challenges such as pressure on infrastructure and the capacity to maintain and develop critical infrastructure as the city develops beyond its capacity. Furthermore, equality and poverty seem to occupy many cities – from the most to the least developed. This is further linked to ideas of community involvement, the establishment of safe neighbourhoods, the issue of unstable labour markets, and fundamental concerns about welfare and social well-being.

Despite the overall overlap of core agendas across cities, the document analysis also reveals that the consensus only goes so far. The resilience strategies are not alike when I compared concrete initiatives and projects. For example, the majority of the strategies reflect a wish to change by turning challenges into opportunities, but they have very different perceptions of what 'change' is. Some cities want to maintain their current municipal functioning despite disturbances, while others want to transform basic socioeconomic structures and re-negotiate the relationship between government and citizens.

Analysing the 100RC toolkit

Having reviewed the main tools developed to support the consultancy dimension of the 100RC, I have created a foundation for discussing the 'softness' of the 100RC toolkit. By applying the analytical framework (membership obligation, steering logic, and interpretive flexibility), I seek to broaden the understanding of how the 100RC makes members comply with network goals.

Membership obligation

All of the three consultancy tools (Resilience Framework/Strategy Process, CRO, and Resilience Strategy) are obligatory for member cities, and each of the tools must be adopted shortly after the city becomes a member of the 100RC. The consultancy tools are thus all mandatory membership obligations and have a high degree of 'hardness' associated with them. Despite the fact that membership in the 100RC is voluntary, there are considerable costs associated with leaving the network. The network offers both direct and indirect financial support to its members, which is immediately lost if the city leaves the network. During one of my conversations with one of the CROs, he told me a story of another city. This city was originally selected for membership in the 100RC and, as part of the induction phase, a team from the network secretariat visited the city in question to meet with the city officials and politicians in charge of the 100RC membership. The 100RC team was disappointed with the city's engagement with the network and decided to cancel the city's membership. The important takeaway from this story is not only was a city excluded because it lacked the commitment to the 100RC process, but that the threat of membership exclusion means that current CROs pay careful attention to how they comply with network goals. The 100RC is considered an exclusive network that is costly to be excluded from, and even though membership eviction is rare, the sanction is very real to the cities. Membership exclusion can be understood as a central sanction mechanism that motivates the members to comply with membership obligations.

Steering logic

The steering logic of the tools is to a large extent implemented top-down. All three tools are authoritative in the sense that they are developed and framed by the 100RC and 100RC partners and subsequently introduced in member cities. The tools reflect the steering logic of a benign top-down approach where few actors have been responsible for the shape, content, and methodology of the tool implementation process. This authoritative steering logic is present in everything from the design of the CRO position to the Resilience Strategy. My interviewees from the 100RC regional offices focused on the CRO and the importance of appointing the right person to do the job in each city. As one of my interviewees put it: 'We have a huge focus on getting the right person. It needs to be the right person that is well connected and close to the mayor' (Interview, Regional Director, 100RC). The CRO is specifically hired on her presumed capacity to carry out the Strategy Process and on her capacity to connect and garner support from local stakeholders and political leadership.

In one of my many conversations with the CRO in one of the cities, the CRO said that he sometimes finds the steering logic too hard and the top-down framing of the Strategy Process too rigid and unrealistic. The 100RC organisation itself is aware of how this authoritarian logic sometimes creates tension in the cities:

> It is not easy to use the same tool everywhere in the world, but it is a common base to identify shocks and stresses. It is a North American organisation, so cities in the US are more used to these types of tools and processes.
>
> (Interview, Regional Director, 100RC)

On the other hand, the 100RC perceives these tools as essential for ensuring that cities comply with network rules and steer towards network goals. When I ask one of the directors why the 100RC insists on deploying a similar toolset across all cities, he answered that it is a way 'to make sure that it happens. And that it happens within a certain time frame' (Interview, Regional Director, 100RC).

Tool flexibility

All the consultancy tools are flexible to some extent. So where the degree of obligation to adapt network tools and the steering logic reassemble what we in governance literature understand as 'hard measures', the ability to design and adapt the tools to a local context is high. The only slight exception to this flexibility is the CRO, whose job description is defined to some extent by the

100RC. That said, the flexibility to translate and adapt both the Resilience Framework and the Resilience Strategy to a local context is very pronounced, which is also illustrated in my review of the resilience strategies. This flexibility is a deliberate choice by the 100RC organisation, which acknowledges the need for cities to make the resilience agenda their own:

> REGIONAL DIRECTOR: The tools are a basis to help. But it is good for the city to have capacity to adapt.
>
> ME: So you don't insist on a particular application of the tools?
>
> REGIONAL DIRECTOR: No, we try not. But as you can see, we have strategies more than 400 pages [long] and we have strategies that are less than a 100. So it is up to the local culture, the local reality. The main objective is to create resilience and to change, but the way they do it is of course different.
>
> (Interview, Regional Director, 100RC)

In the city of Vejle, this flexibility of the provided tools is what makes the 100RC framework attractive and purposeful. One the one hand, the cities are given tangible and concrete tools with the specific purpose of kick-starting a policy formulation dialogue on resilience building. On the other hand, the cities are allowed to contextualise this dialogue according to what the daily life in a given city is all about. A civil servant at the municipality of Vejle commented on the flexibility of the resilience tools:

> It is an important part of the whole process. And I do not know if it was designed by the 100RC with this intention, but if it was, it is a clever design. In this way, it becomes a part of the process to discuss what resilience actually is. Finding out what it means for the city.
>
> (Interview, Vejle Municipality)

The analysis of the 100RC network is summarised in Table 13.2, in which I combine the selected tools with the three characteristics of the governance tools as presented in the analytical framework.

Our understanding of TMN governance becomes more nuanced when we look at obligation, the degree of hierarchy in the steering logic and at the flexibility of the tools. As with most tables, the results appear stronger than the complexity of the reality it seeks to reflect. It is important to note that the

Table 13.2 The 100RC's governance tools

Tool	Obligation	Hierarchy	Fixedness
The CRO	High	High	Medium
The Resilience Framework/ Strategy Process	High	High	Low
Resilience Strategy	High	Medium	Low

different hard/soft measures are not easily categorised and that none of the tools are strictly hard or soft. It would make equal sense to measure the degree of obligation, hierarchy, and fixedness on a continuum. The table, however, serves the purpose of illustrating the main findings of applying the developed framework to the governance tools. Because the key governance tools offered by 100RC are developed from the top-down and framed by the 100RC's central organisation, they are hierarchical and authoritative in nature. On the other hand, the tools allow for local interpretation and application of the tool, which allows different political actors with different contextual constraints to share similar visions and language on how to act on a changing world. In other words, they are not very fixed.

The framework's application to three main 100RC tools reveals that TMN governance goes beyond strictly soft measures and that central 100RC actors firmly guide member cities towards compliance with network goals. It appears that the absence of the traditional toolkit of national states should not restrict us from analysing TMN governance outside the box of soft governance.

Conclusion

The popularity of transnational municipal networks has increased substantially in recent decades. We continue, however, to have limited understanding of the governance mechanisms put in place to steer member cities towards central rules and network goals. In this chapter, I have proposed an analytical framework that helps increase our understanding of the governance tools put in place by TMNs, which help us understand the enforcement structures and embedded positions of authority between city and network. I find that the governance tools supporting the network's consultancy dimension are quite 'hard' in the sense of membership obligations and steering logic. Cities are obliged to engage actively in network activities and adopt a range of tools designed by 100RC and 100RC partners. The tools are very 'soft' in the sense that the 100RC allows a high degree of flexibility in the local adaptation and interpretation of them. Nevertheless, the analysis shows that by imposing membership obligation (and sanctions if not complied with) and by developing and providing certain content to central network tools, TMNs become powerful actors with the capacity to enforce membership rules among city members.

Despite its being an in-depth study of a single network, this case of 100RC opens up a new set of important questions that generally relate to all TMNs. First, it questions the diversity found in the landscape of transnational city networking. The early studies of TMNs claiming explanatory power to polycentric Type II governance are based on networks with seemingly different governance characteristics than the 100RC. We need deeper knowledge about the character of governance tools developed and implemented across different networks in order to better understand the diverse actors operating under the umbrella of the TMN concept. The analytical framework presented in this chapter may be a first step in conducting this exercise.

These nuances may raise further questions of the efficiency and legitimacy of TMN governance. Are networks with 'harder' toolkits more efficient in terms of reaching network goals? And do networks with less polycentric decision-making structures create problems of legitimacy? If TMNs are powerful political actors with tools and ambitions to change local policies, then it becomes important to question whose and which interests are being promoted. The 100RC is not a value-neutral actor, it is a political actor whose worldviews are channelled into active political strategies through its different governance tools. TMN membership moves politics to a new transnational arena about which we continue to know very little.

Acknowledgements

I would like to thank Jens Hoff, Peter Newell and Maryam Hariri for valuable comments on earlier versions of this chapter.

References

100 Resilient Cities (2018) 100 Resilient Cities. Available at: www.100resilientcities. org/partners/ (accessed 16 May 2018).

Abbott, K.W. and Snidal, D. (2000) Hard and soft law in international governance. *International Organization*, 54(3), 421–456.

Abbott, K.W., Keohane, R.O., Moravcsik, A., Slaughter, A.M., and Snidal, D. (2000) The concept of legalization. *International Organization*, 54(3), 401–419.

Acuto, M. (2013) *The Urban Link. Global Cities, Governance and Diplomacy*. Abingdon and New York: Routledge.

Acuto, M. and Rayner, S. (2016) City networks: breaking gridlocks or forging (new) lock-ins? *International Affairs*, 92(5), 1147–1166.

Andonova, L.B. and Tuta, I.A. (2014) Transnational networks and paths to EU environmental compliance: evidence from new member states. *Journal of Common Market Studies*, 52(4), 775–793.

Andonova, L.B., Betsill, M.M., and Bulkeley, H. (2009) Transnational climate governance. *Global Environmental Politics*, 9(2), 52–73.

Barber, B.R. (2013) *If Mayors Ruled the World: Dysfunctional Nations, Rising Cities*. New Haven, CT: Yale University Press.

Betsill, M. and Bulkeley, H. (2006) Cities and the multilevel governance of global climate change. *Global Governance*, 12(2), 141–159.

Blomqvist, P. (2016) Soft and hard governing tools. In C. Ansell, and J. Torfing (eds), *Handbook on Theories of Governance*, Chapter 23. Cheltenham: Edward Elgar Publishing.

Bouteligier, S. (2012) *Cities, Networks, and Global Environmental Governance: Spaces of Innovation, Places of Leadership*. New York: Routledge.

Bulkeley, H. and Newell, P. (2015) *Governing Climate Change*. London: Routledge.

Bulkeley, H. and Kern, K. (2006) Local government and the governing of climate change in Germany and the UK. *Urban Studies*, 43(12), 2237–2259.

Bulkeley, H., Davies, A., Evans, B., Gibbs, D., Kern, K., and Theobald, K. (2003) Environmental governance and transnational municipal networks in Europe. *Journal of Environmental Policy & Planning*, 5(3), 235–254.

Busch, H. (2016) Entangled cities: transnational municipal climate networks and urban governance. PhD thesis, Lund University, Sweden.

Bowen, G.A. (2009) Document analysis as a qualitative research method. *Qualitative Research Journal*, 9(2), 27–40.

Flyvbjerg, B. (2001) *Making Social Science Matter: Why Social Inquiry Fails and How it Can Succeed Again*. Cambridge, UK: Cambridge University Press.

Green, J.F. (2017) The strength of weakness: pseudo-clubs in the climate regime. *Climatic Change*, 144(1), 41–52.

Gupta, R., Pfeffer, K., Verrest, H., and Ros-Tonen, M. (2015) *Geographies of Urban Governance: Advanced Theories, Methods and Practices*. New York: Springer.

Hoff, J. (2018) The role of civil society actors in climate change adaptation strategies: the case of New York City. In T. Scavenius, and S. Rayner (eds), *Institutional Capacity for Climate Change Response: A New Approach to Climate Politics*. London: Routledge

Hoff, J. and Gausset, Q. (2015) *Community Governance and Citizen-Driven Initiatives in Climate Change Mitigation*. London: Routledge/Earthscan.

Hooghe, L. and Marks, G. (1996) Europe with the regions: channels of regional representation in the European Union. *Publius: The Journal of Federalism*, 26(1), 73–92.

Hooghe, L. and Marks, G. (2001) *Multi-level Governance and European Integration*. Lanham: Rowman & Littlefield.

Keiner, M. and Kim, A. (2007) Transnational city networks for sustainability. *European Planning Studies*, 15(10), 1369–1395.

Kern, K. and Bulkeley, H. (2009) Cities, Europeanization and multi-level governance: governing climate change through Transnational Municipal Networks. *Journal of Common Market Studies*, 47(2), 309–332.

Kirton, J.J. and Trebilcock, M.J. (2017) *Hard Choices, Soft Law: Voluntary Standards in Global Trade, Environment and Social Governance*. London: Routledge.

Lee, T. and Jung, H.Y. (2018) Mapping city-to-city networks for climate change action: geographic bases, link modalities, functions, and activity. *Journal of Cleaner Production*, 182, 96–104.

Nielsen, A.B. (2016) Diplomati i en urban tidsalder. *Økonomi & Politik*, 89(4).

O'Leary, Z. (2014) *The Essential Guide to Doing Your Research Project*. Thousand Oaks: SAGE Publications.

Rashidi, K. and Patt, A. (2018) Subsistence over symbolism: the role of transnational municipal networks on cities' climate policy innovation and adoption. *Mitigation and Adaptation Strategies for Global Change*, 23(4), 507–523.

Rayner, S. (2010) How to eat an elephant: a bottom-up approach to climate policy. *Climate Policy*, 10, 615–621.

Shaffer, G.C. and Pollack, M.A. (2009) Hard vs. soft law: alternatives, complements, and antagonists in international governance. *Minnesota Law Review*, 94.

Tyler, S. and Moench, M. (2012) A framework for urban resilience. *Climate and Development*, 4, 311–326.

14 Sustainability, democracy and the techno-human future

Clark A. Miller

Introduction

The starting point for sustainability is generally taken to be human–nature relations: do human affairs fall within planetary boundaries (Young *et al.* 2006; Rockström *et al.* 2009)? I believe this is deeply mistaken. In my judgement, human–technology relations are far more central to sustainability. Humans today are hybrids with their technologies: they are techno-humans who design, inhabit, comprise, and become products of the vast socio-technological systems out of which societies are made and that define and organise human–nature and human–human relations. The behaviour of these systems – energy, food, water, transportation, communication, manufacturing, the built environment – drives unsustainability. Creating a sustainable human future requires transforming these systems, that is, the ways that people design and use technologies and the reciprocal implications for social practices and organisation and what it means to be human. Sustainability must thus begin with an analysis of human–technology relations and the capacity to govern them: democracy among techno-humans or the self-governance of socio-technological systems (Allenby and Sarewitz 2011; Harari 2016; Frischmann and Selinger 2018).

We inhabit a world created and organised by the power of science and technology to fashion every aspect of modern societies (Jasanoff 2016). Consider the biological make-up of the human population in the twenty-first century. Our biology no longer reflects a prehistoric or God-given human nature. The collective composition of human bodies on the planet is an outcome of the nutritional content of the food that societies manufacture, distribute, and eat; the pharmaceutical chemicals they ingest into their bloodstreams; the technological environments that distribute injuries and the chemicals and infectious diseases to which bodies are exposed; the health care systems that determine who is treated for which illnesses and in what ways; the technologies that enhance, augment, supplant, or regulate reproduction; the implications of military technologies for the physical and mental experiences of soldiers and civilians; the sciences that shape the understanding and practice of these diverse systems; and the laws, norms, markets and forms of communication, and individual and collective decision-making that govern them all (Hacking 1990; Haraway 2000).

Human biology is literally a product of the ways that socio-technological systems (Hughes 1983) make and metabolise food, energy, water and materials and construct and operate agriculture, electricity, transportation, communication, manufacturing, computing, and the built environment. So, too, are human ideas, which circulate among the telecommunications networks that comprise the news, social media, the internet, telephones, etc. We are what we read, Benedict Anderson suggested about nationalist imaginaries shaped by the technology of newspapers and novels, the principle recent change being the growing organisation of reading practices via social media (Anderson 1991).

Socio-technological systems, I argue, are the central drivers – the root cause – of planetary unsustainability. They structure social relationships and practices of imagining, knowing, narrating, deciding, acting, and making (Shove and Walker 2014). In turn, they link human decisions and actions to global patterns of social and ecological footprints (Busch 2011). It is no accident that today's discussions of sustainability focus on the pipelines and mines of Exxon Mobil and the Koch industries while discussions of democracy and inequality focus on the networks and algorithms of Facebook, Google and Apple. The planetary-scale socio-technological systems managed by these and other companies – far more immediately and directly than nations, ideologies, or pools of capital – constitute human organisation and experience, distribute power and wealth, and transform nature, biology and ecology in the twenty-first century (Mitchell 2011). It is these systems that we must learn to govern effectively if we are to meaningfully create sustainable human futures.

In this chapter, I explore the question of what it would take to address the crisis of self-governance of socio-technological systems at the heart of global unsustainability. Can techno-humans govern the systems they have created for, and in which they have embedded, themselves? I explore this question in four parts. First, I reiterate in somewhat greater depth and specificity the proposition that unsustainability should be ascribed, fundamentally, to socio-technological systems and that, as a consequence, to pursue sustainability requires a much more explicit focus of the critical gaze on the organisation and operation of these systems. Second, I suggest that our current ways of imagining and organising democratic governance are badly misaligned with human relations with technology. This misalignment is bound up in the basic epistemic and political organisation of the state and its capacity to know and thus govern socio-technological systems. Third, I argue that the solution to this misalignment is to co-produce new forms of knowledge and governance focused explicitly on socio-technological systems. Co-production has traditionally been understood as a tool of critical analysis, but I suggest that it can also be retrofitted as a design science for the re-imagination and reconstruction of socio-technological systems and their forms of governance.

Finally, the chapter concludes by arguing that a new focus on co-production as design will highlight the valuable role for empowered communities in advancing sustainability. To make that case requires rethinking community among techno-humans. If people are techno-human, so too must be communities.

Communities in the twenty-first century are no longer populations primarily defined by geography – the proverbial next-door neighbours. Communities, I suggest, co-extend in space and time with socio-technological systems, some of which are geographically defined but many of which are not. Communities, in other words, comprise the diverse inhabitants of socio-technological systems – and thus are the heart of what it would mean to construct adequate forms of socio-technological self-governance. To the extent we govern our own socio-technological inhabitances as socio-technically networked communities across the full complexities and reaches of socio-technological systems, then it is possible to say that we have a working democracy. To the extent that we do not, then we will struggle to confront unsustainability – or other grand challenges in global societies, such as persistent and deep inequalities – because, ultimately, we will fail to hold ourselves, our institutions, and our relations with technology accountable for their global social and ecological footprints.

Unsustainable systems

Mary Shelley's *Frankenstein*, which celebrated its 200th anniversary in 2018 (Shelley *et al.* 2017), often serves as the lens through which to understand techno-human relations and, especially, the moral and ethical dilemmas of technology: technological hubris and technology's escape from human control to wreak destruction on people and the environment. *Frankenstein*'s narrative elements reverberate through the histories of innovation and its critics. Yet the story's emphasis on the solitary inventive genius of Dr Victor Frankenstein and on the isolated technological subject/object of the monster – so evocative of the discourses that surround technologists (from Thomas Edison and Henry Ford to Elon Musk, Steve Jobs and Bill Gates) and technologies (the light bulb, Model T, Tesla Model X, iPhone, PC computer) – in fact badly distorts our perceptions of the challenges that technology poses for societies.

For problems such as sustainability, individual devices or apps matter less than the embedded, networked, globalising systems that have come to permeate technological societies; individual inventors matter less than the armies of engineers that design, build, operate, and maintain those systems; stories such as *Frankenstein* matter less than more modern prophecies, such as George Orwell's *1984* and Aldous Huxley's *Brave New World*, that better capture the power and pathologies of what urban theorist Lewis Mumford called, in his 1930s masterpiece *Technics and Civilization*, 'The Machine': the collective orchestration of human and machine in the service of organised purpose (Mumford 1934).

Industrialisation – whose mechanisation not only of production but also of life captures the imagery of so many modern adaptations of *Frankenstein*, but was largely absent from Shelley's original novel – has transformed the imagination of human–technology relations far beyond Shelley's dreams of the individual techno-human gone awry. Beginning in the second half of the nineteenth century, the deployment of machines began to transform work and transportation in a widespread way, bringing into being not only factories, railroads, and the production and refining of oil and coal but also new ways of understanding time, managing human relations, and conceptualising what it means to be human (Galambos 1970; Chandler 1977; Noble 1979). Science itself became no longer the work of individuals but of massive enterprises. By the middle of the twentieth century, one such enterprise was capable, in a few short years, of creating an extensive network of new scientific laboratories devoted to nuclear research at Los Alamos, Livermore, Chicago, Oak Ridge, and Hanford, of inventing the atomic bomb and producing tens of thousands of such weapons in the world's largest industrial plant at the Clinton Engineer Works and in manufacturing facilities across the United States. A quarter of a century before, in 1919, US automobile manufacturing had already reached one million vehicles per year, creating the impetus for cities to rapidly expand outward and to become places designed less around people than cars, ordered more than anything by avenues of concrete for driving from home to work. Cities in the United States were largely electrified by the 1920s; the rural countryside by the 1950s. The 1956 US Federal Highway Act launched the construction of a network of massive arterial freeways to enable cars and, increasingly, transportation trucks to traverse the nation. Along the way, humanity transformed itself into both machines – modern medicine views, and acts on, people as though they are mechanico-chemical hybrids, laying both the conceptual and material foundations for human performance enhancement – and cogs in the machine of an industrial and technological workforce. The economy is now measured in terms of its productivity and employment; our worth as individuals measured by our contribution to the making and operating of the vast machinery of modern societies.

It is these socio-technological systems that drive unsustainability. Each links together ideas about what societies are, how they work, and how people fit into them; social networks, relationships, and organisations; and technological capabilities, in ways that have now evolved far beyond the imaginations of their early advocates and instigators. The systems are not just at odds with Earth systems, they are integral to them. Just as it no longer makes sense to approach public health as a problem of human biology independently from the technological systems and environments within which human bodies reside, so it no longer makes sense to imagine or study ecology independently from the techno-human (McKibben 1989). At all scales, from the microbial to the planetary, Earth systems now revolve around technology as much as the Earth revolves around the Sun. The products of industrial, pharmaco-chemical, and material transformation now drive rapid changes in bacterial and viral speciation (e.g. through reactions to antibiotics and through the integration of human-avian-pig ecologies in diverse farming systems); animal and human nutrition (e.g. through plant and animal breeding and the industrialisation and later genetic engineering of agriculture and food science); and the nitrogen, phosphorous, and carbon cycles and the climate (e.g. through the burning of fossil fuels and the production of artificial fertilizers). Much the same can be said of democracy and the economy: neither politics nor markets now exist (if they ever did) locally or globally, independently from the technological organisation and mediation of money and power, desire, interest, the human body, or our social relationships (Ezrahi 1997; Carmen 2004; MacKenzie 2008).

Take carbon as an illustrative example (see, for example, Pontifical Council for Justice and Peace 2014). The material economy of carbon – vast systems for mining, pumping, transporting, refining, and burning hydrocarbon fuels – generated companies that were the largest in the world for well over a century. Their power and influence shaped political economies in every corner of the globe, enabling not only manufacturing, industry, petrochemicals, and automobiles but also the enormous armies of the twentieth century. The supply chains to create this abundance became, for the imperial powers of Europe and North America, the global raison d'être of military security, often defining when and where the great powers deployed their armies, fought wars, and carved up colonial territories, pre- and post-independence. Carbon-based supply chains distributed wealth to the elite few, impoverishing and destroying the health of others, arguably driving inequality and pollution as much or more than any other industry in human history (Perkins 2011). Carbon also transformed our ideas of democracy, culture, and consumer societies, rooted in inexpensive, abundant sources of fossil fuel energy (LeMenager 2014).

Carbon economies directly drive all three of the existential global environmental crises currently confronting humanity: *climate change* via the direct injection of methane and carbon dioxide into the Earth's atmospheric systems; *biodiversity loss* via the application of oil-powered industrial machinery (tractors, chainsaws, etc.) to the destructive transformation of the world's forests and grasslands into agricultural fields and pastures; and *the depletion of the ozone layer* via the petrochemical manufacture and release into the atmosphere of diverse species of chlorofluorocarbons – a class of chlorinated and fluorinated hydrocarbons. At the other end of the problem, climatic changes now being driven by atmospheric concentrations of carbon dioxide and methane create risks for human societies precisely because they disrupt the powerful socio-technological infrastructures that human societies have come to rely on for security: agricultural, urban, manufacturing, and military systems that provide food and water, protect against the weather, create a robust economy and jobs, and defend against threats of violence.

Can humanity exert control over our own techno-human futures through effective practices of what we might call socio-technological self-governance? This, to me, is the central question of sustainability in the twenty-first century. The question haunted Mumford, who asked whether humanity could manage The Machine – the orchestrating technological organisation of human life and work. Mumford saw what has remained foggy for many up to now, that is, the degree to which what is human in biomaterial and organisational terms is now thoroughly defined by the mechanical systems that order our societies. Each of these authors despaired deeply about the trajectories of modern societies toward pathological forms of social mechanisation and the degree to which societies seemed blind to those consequences. A half-century after Mumford's work, techno-political theorist Langdon Winner named this neglect *technological som-nambulism*: sleep-walking through extensive transformations of what he termed the technological constitution of politics, political economies, and the ordinary practicalities of daily life (Winner 1986). It would seem our existing governance mechanisms are poorly fit for the purpose of socio-technological self-governance.

Governance failures

Why do our current practices of democratic self-governance fail so miserably to address the global sustainability crisis – and other grand challenges, such as global inequality or the resilience of societies and infrastructures to emerging climatic risks – grounded in the technological constitution of modern societies? Many blame capitalism, or variants of capitalism, e.g. neoliberalism or consumerism, and the corruption of market logics (Klein 2015). Others see today's challenges as the result of persistent histories of colonial and/or authoritarian politics (Agarwal and Narain 1991). Still others emphasise the need for a new science or ethics of sustainability in individual and institutional decision-making (Thompson 2010). I offer a different diagnosis: our current democratic self-governance practices fail because of a mismatch between the ways that democratic societies have come to imagine and practice governance over the past two centuries and the organisation of socio-technological systems.

The problem of sustainability is not, per se, that individuals or organisations make choices to act or to buy things in the service of diverse interests, goals, or ideals. It is instead the ways in which those choices are structured by socio-technological systems and the rippling rivulets of transformation that trickle outwards from those choices through intertwined socio-technological systems to create patterns of social and ecological footprints. When someone buys an iPhone, people somewhere have dug its material components out of the ground, processed and manufactured them into the iPhone by using designs from others in other places, and shipped the device around the globe to be sold. The use of the iPhone then interpolates and interpenetrates other relationships, enabling new ways of living, working, and engaging for its user and those with whom her actions interact and intersect. Yet, this web of interrelationships and inter-actions is largely invisible. We know they are there. Even the most sophistic-ated analytical tools of sustainability, however, such as lifecycle analysis, provide only the merest glimpses into their extent and complexity.

The problem of knowledge – or self-knowledge – within socio-technological systems is central to the sustainability crisis. Sustainability science has long recognised this challenge, but in my view has focused on the wrong aspects: emphasising knowledge of human–nature relations and viewing it as a problem of knowledge alone (Kates *et al.* 2001). I suggest the problem lies deeper – in how the state and its knowledge systems are organised together.

Recent decades have witnessed a fundamental transformation in the scholar-ship of knowledge and governance. At the heart of this critical reassessment lies the idea of co-production: the bi-directional relationships among the ways that human societies know and order their worlds (Jasanoff 2004).[1] What we know is a product of socially ordered practices of knowing: the aspects of the world people and institutions choose to observe and theorise, the data they collect, the methods of analysis and reflection and epistemological and ontological assumptions they adopt, and the norms and standards of evidence they apply. Simultaneously, what and how people know fundamentally shapes their interpretations of the people and events around them; their identification, analysis, and selection of responses; and the organisation they imagine and impose on society, nature, and technology.

I suggest that the crisis of unsustainability is wrapped up in the specific ways of knowing and governing imagined and practised by contemporary democratic societies. For sustainability purposes, the state and its knowledge systems are organised wrongly. We do not know what we need to know about socio-technological systems to effectively govern them because democracy *as currently organised* neither creates nor is positioned to make use of such knowledge.

Michel Foucault's notion of *governmentality* offers a starting point (Foucault 1991). For Foucault, the eighteenth- and nineteenth-century ideal of government was invented to encompass the rational management of society. In this ideal, the state's power and authority to police its subjects was grounded in the intertwining of several elements: the idea of society as a population that behaved according to knowable laws; the idea of the improved welfare of the population as the purpose of government; the sciences of population (e.g. statistics) as the forms of state knowledge underpinning actions taken to improve public welfare; the idea that the population is also the object of government action; and the specific apparatuses of governance through which the state acted upon society, informed by the sciences of population, in order to improve social welfare.

Building on Foucault's ideas, numerous scholars have explored the multiple 'institutional transformation' during this period, especially in the second half of the nineteenth century (Wittrock and Wagner 1996). Emerging through dynamic interactions, numerous new institutions appeared, including new government agencies devoted to the rational management of the welfare of populations (across diverse fields from labour and social work to mental health and criminal justice); the forms of knowledge (statistics of poverty, health, economy, etc.) necessary to support and justify managerial decisions; the agencies that collected and curated relevant statistical data; the trained managers who staffed these agencies; the professional societies they belonged to; the schools that trained them; the forms of professional licensing that governed the exercise of their disciplinary powers; the modern research university; and the social science disciplines and research fields, methods, and institutions that defined key statistical quantities and used them to theorise and analyse both the welfare of populations and the government actions taken to improve this welfare (Nowotny 1990; Porter 1996; Rueschemeyer and Skocpol 1996). The result was a complex, layered co-production of diverse forms of knowing and ordering within the administrative state, epistemic and disciplinary institutions, and an increasingly normalised society.

In turn, these same ideas became influential in shaping democracy. The broad socialisation of ideas of population and the rational pursuit of public policy among democratic publics has had a profound impact on the democratic imagination (Ezrahi 2012). The politically self-aware imagination of the nation as a spatially bounded, sovereign people – a population with sole legitimate political jurisdiction and authority over the territory it inhabited – spurred multiple successive waves of efforts to end both colonial and imperial rule around the globe.[2] Indeed, nationalism continues today to fuel robust democratic ambitions among populations seeking self-governance, including widespread indigenous nations around the globe.

The welfare of the population – a notion captured, e.g. in the idea of the greatest good for the greatest number – similarly contributed in the early twentieth century to the democratisation of the benefits of state policy, both in the sphere of social policy (Skocpol 1992; Rabinbach 1996) and natural resources (Hays 1959). Following the Second World War, similar logics became the foundation for a new politics of innovation that justified government investment in science on the basis of the social and economic benefits that would flow to health, welfare, economy, and security, further reinforcing the commitment of the state to a technological model of economic development (Bush 1945). Underpinning all of the above was the imagination of expertise, rationality, and quantification as instruments for depoliticising the exercise of political power, thus rendering the latter more legitimate in the eyes of democratic publics wary of the arbitrary and capricious actions of state authorities (Ezrahi 1990; Jasanoff 1990; Porter 1996). These modalities not only rendered nature and society legible to the state (Scott 1998) but state actions visible and transparent to citizens (Eschenfelder and Miller 2007; Ezrahi 2012).

Drawing all of this together, it is possible to begin to diagnose why socio-technological self-governance has failed so spectacularly in dealing with the emergent challenges of unsustainability. The foundational knowledge systems of modern democracies – the statistical and scientific agencies and datasets fashioned to feed the insights of the state and its publics into the world they inhabit – remain largely social and natural: they are rarely technological and even less so socio-technological. Where they concern the (technological) economy, they favour measurements of inputs (resources), outputs (products), workforces (labour), and productivity to detailed mappings of complex dynamics and networks that comprise modern socio-technological systems or the ways in which these systems constitute diverse forms of people, livelihoods, communities, and political economies across their distributed geographies. These measurements thus illuminate inequality (e.g. Gini coefficients) and unsustainability (e.g. anthropogenic carbon emissions) but fail to make visible how and why socio-technological systems cause these outcomes or what might be done to alter them.

Contemporary democracies also confront an epistemic mismatch between national geographies and global socio-technological systems. Global technological infrastructures transcend the limits of any nation and facilitate the global movement of people, goods, and services. Facebook has over one billion inhabitants of its socio-technological networks, spread across all countries. Its media presence has become a conduit for foreign influence and interference in the domestic politics of democratic societies. The oil industry has similarly corrupted local politics across the globe, building global technological infrastructures capable of supplying the voracious thirst of economies everywhere for liquid fuels for equipment, automobiles, aeroplanes, and factories. Yet, the knowledge systems of democratic governance largely remain domestic in focus. Only very recently have serious efforts been made to extend either knowledge systems or logics of governmentality to global affairs, with limited success, and such efforts are now facing a severe backlash around the globe (Miller 2004, 2015).

Last, but certainly not least among the core failures of socio-technological self-governance, is the state's commitment to the expansion of socio-technological systems as instruments not only of biopolitical governmentality (e.g. the police, health systems) but also of techno-economic and military success. As the direct owner of the means of the production, as regulator, and as financier of infrastructure and innovation, the state and democratic societies have bound themselves to imaginaries of technological progress as the foundation of societal success and population welfare (Jasanoff and Kim 2015). The success of these visions has tended to draw attention away from the sacrifices that technological development has required of societies, especially – but certainly not exclusively – where those have conveniently lain beyond their sovereign borders.

Contemporary forms of governmentality have simply not required – indeed have actively discouraged – seeing the socio-technological constitution of the world. States today are hardly ignorant of the downsides of technologies. Tort law, risk assessment, consumer and environmental protection, and safety and efficacy requirements for new drugs all reflect awareness of technological harm (Jasanoff 1996). Yet, the harms that have been successfully excised from techno-economies of production are broadly ancillary to the operation of socio-technological systems. Unsustainability has proven much more challenging: the biological hazards of daily chemical ingestion by billions of people; the permeation of our lives and ecosystems by plastics; the production of carbon dioxide in energy systems; the waste streams of factories; the sacrifice zones of supply chains and their hazardous work and environmental conditions; the segregation and mechanisation of workforces; the continuing upward spiral of destructive military force; and the extensive collection and algorithmic processing of data by Facebook and other technology companies. These are not mere by-products but rather core, built-in functionalities of socio-technological systems. And in these functionalities, more than just harm is at stake. What will it mean to be human? What kinds of lives and work will people pursue? What kinds of bodies and minds will they have? On these kinds of questions, existing forms of socio-technological self-governance are largely silent, except insofar as they uncritically adopt the hyped narratives of techno-economic self-promotion.

Redesigning knowledge for sustainability

Can we do better? I believe so. The grand challenge of the twenty-first century is to reconfigure the world's socio-technological systems to create more sustainable and equitable human futures. This exercise entails redesigning both the knowledge systems through which societies produce insights into their socio-technological selves and societies' capacities to deploy those insights for the purposes of governance. It is a daunting call to rework governmentality for the twenty-first century, that is, to reimagine populations, societies, the social sciences, and government around socio-technological systems. It is thus a call to retrofit the co-production of knowledge and governance as a design science.

I suggest this exercise begins with the reorientation of the human sciences. Research in science and technology studies (STS) has provided a rich foundation of theoretical insights into the complex constitutions of techno-humans and the socio-technological systems within which they reside (Felt *et al.* 2016). With thousands of practitioners and dozens of PhD training programmes around the globe, STS has reinvented humanistic and social scientific analytics to produce vibrant and finely textured analyses of science and technology as human institutions, wrapped up in and transforming every facet of human societies: politics, markets, militaries, families, the law, etc. Through this work, STS has provided the foundation on which to build new techno-human sciences for sustainable socio-technological societies.

Additional steps remain, however. We need much better tools for mapping, measuring, monitoring, and modelling global socio-technological systems. Opportunities exist to exploit large-scale data collection and synthesis, systems and network modelling, geographic information systems, and other research tools to provide real-time, comprehensive pictures of socio-technological systems in ways that link the ability of STS to reveal the micro-social realities of the techno-human constitution to the macro-social, economic, and political dynamics and structures of global food, energy, manufacturing, and communications systems. In energy transitions research, for example, new approaches are needed to visualise socio-energy systems (Miller *et al.* 2015), to examine how such systems enrol and construct people, and to explore what the transformation of such systems might mean for the futures of diverse groups and communities across the planet. For example, no map exists of the socio-energy system comprising the global manufacturing, supply chains, and disposal of photovoltaic technologies, with this system's human and material geographies, and the human dimensions of increasing the size of that enterprise by an order of magnitude in future decades.

In another illustrative example, as my colleague David Guston has observed, despite the fact that the power and wealth of companies such as Facebook, Apple, Google, Microsoft, Monsanto, and Exxon Mobil rival that of many nations, Google Studies doesn't exist as a coherent subject, let alone as the organising principle of modern human sciences. The narrative forms of research produced by STS scholarship are incredibly powerful, especially when coupled with ethnographic methods and the humanistic capacity for exploring the construction and articulation of meaning in human societies that is all but absent from more quantitative social sciences. But these approaches are also insufficient. A rewrite of the knowledge of socio-technological systems in the human sciences must begin with these foundations but also find ways to tackle vastly greater complexity and scales of analysis to underpin a realistic push for a more sustainable future (Smith and Stirling 2008; Leach *et al.* 2010).

Expanding the capabilities of the human sciences to analyse socio-technological systems will require, in turn, synthesising and reimagining the disciplinary landscape of universities – including the breaking down of disciplinary barriers within the humanities and social sciences, where theoretical research remains bound up with the problems of human populations and the methods of studying them envisioned by nineteenth century models of governmentality – and reintegrating the human sciences and their technological neighbours (Miller 2010). The goal is, at once, to reframe analyses of *homo-economicus* or *homo-politicus* – and of markets, firms, governments, individuals, families, militaries, etc. – in fully technological terms and also to end the pretension, which is endemic in the engineering disciplines, that it is possible to design and analyse technology solely in technical terms (Herkert 1990). The point is not, of course, to end the differentiation of education and research within universities and research centres but to rework the core assumptions on which that differentiation is built so that engineers acknowledge and engage with the reality that they are building human as well as technological worlds and that the human sciences engage with the reality that their subjects are thoroughly techno-human as a result.

The same kinds of changes need to occur within the knowledge systems of the state. The state expends vast resources on knowledge for decision-making, divided, just as in universities, between the social – e.g. demographic, social, and economic statistics – and the technical – e.g. environmental impact and energy production. The state's capacity to finance knowledge-making is unrivalled, even by entities such as Google and Facebook, although the latter two for the first time offer insights into individuals that may rival in depth and sophistication the knowledge of many states. Yet this capacity has largely been squandered with respect to socio-technological transformations within societies. For example, the latest scramble to understand the future of work in technological societies suggests just how deeply the imagination of the state has been captured by the socio-intellectual geographies of the late nineteenth century and its focus on the statistics of populations – in this case, in terms of measures of employment and income, not the kinds and forms of employment characteristic of diverse techno-enterprises and the ways those shift, over time, the nature and organisation of work (World Economic Forum 2016).

To transform the ways the universities and the state know society requires the human sciences to re-emerge from the theoretical cocoon of the past half-century and re-engage as pragmatic sciences of governance (Miller 2010). This is a profound epistemic and institutional challenge. Epistemically, new questions must come to the fore. What does it mean, for example, to rethink societies in terms of socio-technological systems and imaginaries, not just as a theoretical exercise but as objects of governance imagination and practices? What are the proper ends of the governance of socio-technological societies: what, in other words, would it mean to for socio-technological societies to perform well or for their inhabitants to live well? How do legal and governance institutions and arrangements need to be rearranged to make possible the legitimate and effective regulation of global socio-technological systems?

Institutionally, the challenge is to do this via engaged practices, working collaboratively across diverse entities in society – engineering professions, government agencies, corporations, legal institutions – to develop applied insights about the socio-technological foundations of contemporary societies and to put them to work in practice (Downey and Zuiderent-Jerak 2016). Far more emphasis, for example, needs to be placed on preparing professionals to incorporate insights into socio-technological systems in the work of governance. Across the globe, universities train the leadership of legal, administrative, managerial, engineering, medical, and many other professions. Professional education largely ignores the deep integration of the social and the technological in contemporary societies, as well as the insights into these topics offered by STS and socio-technological systems research. Despite the fact that their primary duties will be to oversee the funding, construction, operation, and maintenance of, for example, technological infrastructures, local government practitioners in most public administration professional programmes receive no training in infrastructure studies (Slota and Bowker 2016). Above all, the changes suggested above will require reconfiguring the norms, expectations, and rewards of higher education institutions, enabling theory to be joined by practice and engagement as viable career trajectories for future researchers and educators, and reconfiguring the pedagogical foundations of professional education.

Reimagining and re-empowering community

It is not enough, however, to reinvent knowledge. We must also reinvent governance. The need for new forms of governance to address sustainability is widely recognised, with calls for global governance (Young 1997), fundamental changes in individual behaviour (Vlek and Steg 2007), and everything in between. Arguably most compelling is the increasingly popular idea of distributed governance – variously termed polycentric, patchwork, fragmented, networked, or multi-level governance – involving institutions working together across diverse locations, jurisdictions, scales, and forms of authority (Hooghe *et al.* 2001; Biermann *et al.* 2009; Ostrom 2010; Rayner 2010).

What makes the idea of distributed governance compelling for sustainability is the potential for governance imaginaries and arrangements to be brought into alignment with the complex forms of socio-technological systems that criss-cross scales and geographies from the individual to the planet, from Tipperary to Timbuktu. If contemporary states don't have the right forms of governmentality to govern socio-technological systems, global institutions lack the necessary power and authority, and individuals lack the necessary leverage. New forms of governance will require the ability to follow socio-technological systems wherever they go, from the constitution of their dispersed, techno-human inhabitants to the orchestration of global supply chains to the diverse local and national regulatory regimes within which they operate. Choreographing these diverse institutional and political performances holds the potential to reinvent forms of self-governance explicitly in terms of socio-technological systems. The challenge is to draw together the many varied sites of collective human governance – far transcending government – into a mounting cacophony – or more properly, a symphony (Ezrahi 2012) – of transformational, global action.

What is missing, for me, from current proposals for distributed governance is any real attempt to grapple with the socio-technological foundations of contemporary societies. Like the models of governmentality and democracy from which they have evolved, they retain a deep imaginative divide between the social and the technical. The state and civil society stand apart from technology. To truly reimagine self-governance as socio-technological requires understanding how to re-pattern governance in ways that map onto the forms of socio-technological systems. Not just any polycentric arrangements of governance will do: we need to map the design of distributed governance so that it aligns with the design of socio-technological systems (Voss *et al.* 2006).

What I argue, especially in the context of this volume, is that the empowerment of communities offers an important strategy for pursuing this idea of distributed self-governance of socio-technological systems. I suggest that civil society, with its vast diversity of participant communities, is where the human future will be won or lost. Historically, of course, civil society has been mapped in precisely the same geographic and conceptual terms as population within models of governmentality. Civil society is a property of the nation state. And, yet, the rise of insurgent globalist and localist ways of thinking and organising has begun to profoundly challenge nationalist imaginaries, as has the politics of race and ethnicity, especially with rising immigration. As a global force, civil society is now understood as transcending the nation, encompassing diverse local and transnational social movements, activist organisations, and indigenous communities that increasingly put pressure on state-based models of global environmental governance (Lipschutz and Mayer 1996).

Indeed, if we look below the surface, civil society is not a random ordering of the world into random forms of sociality. It is already well designed for the task of pursuing sustainability because it is broadly organised in close alignment with the socio-technological systems that need to change. Not surprisingly, yet rarely reflected upon, the deeply constitutional role of socio-technological systems in contemporary societies is reflected in the forms of communal life characteristics of modern societies. While many communities are not organised in explicitly socio-technological terms, many others *are* organised in these terms, not only in form but in focus and not only through but in relation to the rising power of socio-technological systems to organise human affairs since the late nineteenth century. Communities of professionals and workers, for example, have frequently organised by (technological) industry, by forms of expertise, and by technological networks of communication and transportation (Layton 1986). The primary organisation of the modern corporation around the operation of discrete socio-technological systems of production (Trist 1981) means that the communities and cultures that form in and among leaders, workforces, customers, critics, regulators, and others often form around particular technological forms of life (see, for example, with respect to the nuclear industry (Hecht 1998; Gusterson 2004). Technologies such as television, the automobile, and the internet have formed powerful bases for community reformation and reorganisation (Turkle 1995; Foster 2003; Hill and Gauntlett 2002). And, arguably the most important form of social life in the twenty-first century, the city – where the vast majority of people now live – is fundamentally socio-technological in organisation (Mumford 1934; Soleri 2012; Taylor and Derudder 2015). We live in thoroughly technological societies, and many of the communal forms of life within those societies already reflect that fact in some fashion or another.

One of the critical features of socio-technological systems is that their governance has been largely shielded from communities over the past century. Dominant modes of governance have focused power in the hands of those who can ensure the technically and economically efficient organisation, operation, and growth of technological systems. The power of diverse communities enmeshed in such systems – e.g. workers, users, non-users, those whose health or environments are impacted by operations – have tended to be downgraded and defused.

Reconfiguring the self-governance of socio-technological systems means identifying the inhabitants of such systems, finding ways to give them new knowledge and new power to shape the future of such systems, and holding them accountable for creating more sustainable and equitable futures. Self-governance is a form of power-sharing, arguably, between and across multiple, diverse forms of knowing and deciding, some of which are individual, some of which involve small groups, and some of which encompass all of society. And the terms of this power-sharing are always under reconsideration and renegotiation. The products are the forms of society and governance, as well as the ways of organising knowing and imagining the thought necessary to guide decisions by the diverse participating elements. Therefore, when we talk about empowering civil society to create new forms of alliances and new knowledges for sustainability and resilience, we are talking specifically about reconfiguring democracy to enable different forms of collectives within society to have different kinds of power and authority to pursue both the knowledge and the decisions or actions they consider necessary to achieve sustainability and resilience.

Humanity has done it before. Human imagination has a powerful capacity to create new ideas, to hone those ideas into knowledges, to deploy those knowledges to organise human relations in new and innovative ways, and then to evaluate whether those new forms of organisation are worth inhabiting and, where necessary, to criticise and ultimately overthrow them. Sustainability is simply the next iteration of that project. Concretely, we are already moving in the right direction. Neighbourhood, urban, or regional governance entities are already co-producing new forms of knowledge and power that enable them to place new demands on technology, such as transportation systems and electricity grids, that are central to their forms of life, e.g. in recent movements by cities and public utility commissions to drive decarbonisation (Bulkeley 2013). Professional communities are actively working to integrate sustainability research and social responsibility into the design of technological systems (NAE 2014). Companies are devising new models of governance that enable tracking the human and ecological footprints of their systems and operations and incorporating the resulting insights into improved decision-making (Busch 2011; Bulkeley and Newell 2015). These are just the tip of the iceberg, however, of what will be required to reorganise the epistemics and politics of democracy as a socio-technological project, something few have imagined at depth. However, I suggest the reader sees Malka Older's impressive *Infomocracy* trilogy for an intriguing glimpse of what such a future of radically innovative knowledge and governance might look like.

In the past 25 years, climate change has gone from an idea understood by a tiny elite to one that is widespread among the world's seven-plus billion people – and widely believed. Jesus and the early Christian church didn't accomplish anything nearly that transformative in anything like that short a time. Socio-technological transformation is slow. Yet, renewable energy is now not only the fastest growing segment of the energy sector, its combined annual additions to the world's energy supply are larger than all other forms of energy combined. Renewable energy is already 1 per cent of the world's energy supply and closer to 3 per cent of the world's final energy use. And from 1 per cent to 100 per cent is a mere seven doublings. At the current pace of acceleration in the renewable energy sector, that may just be fast enough to keep us within the 1.5–2 degrees envelope, even if we don't accelerate either carbon capture technologies or energy efficiency. Change is coming.

Indeed, I am far less concerned with the pace of sustainable transformation than I am its social outcomes. My optimism is the same as my deepest concern. It's not that humanity won't accomplish a sustainability transformation, it's the kinds of futures that we will design to replace our current, unsustainable selves. It is all too easy, hanging with the advocates of artificial intelligence and gene drives, to see a future for humanity in which we are plugged into the sustainable technological systems of the future as digital and biological parts, inhabitants of The Matrix (O'Neil 2017). And it is all too easy, hanging with the advocates of sustainability, to envision its own sacrifice zones (Lerner 2010). Coal mining and the burning of coal are some of the most dangerous, dirty, and unhealthy forms of techno-human construction ever invented. Liberating communities from the coal industry should be an act of emancipation. It is, I think, a deep failure of the human sciences that we have not helped those who design political strategy to understand the failure of imagination inherent in the self-described war on coal. There are many other aspects of the coming transformation where I fear that we fail to fully understand the social consequences of what we are doing as we attempt to transform socio-technological systems towards better futures. In various places in the world, it is already clear that current transformations of energy systems have made those systems less sustainable and resilient, not more, and are creating new sites of sacrifice and new regimes of inequality. Social scientists must do better at understanding the societal ramifications of these techno-economic projects. Self-governance of socio-technological transformation requires a much deeper knowledge of the techno-human present and future.

This only makes more critical the role of the social sciences in the project of sustainability. The societal transformations sought in the name of sustainability are not merely opportunities to make societies carbon-neutral. They are opportunities to pursue great human ambitions for justice, equality, freedom, liberation, emancipation, transcendence, etc. Sustainability is the next iteration of the great fashioning of human societies. Much of what is ill in the world – violence, injustice, inequality – is a direct product of twentieth century socio-technological systems. We have the opportunity to design better, but we cannot do so unless we make social progress an explicit goal of the sustainability project. Ecological friendliness is a nice ambition for lots of reasons, but it isn't sufficient. The social sciences need to insist that the sustainability project be more than just a carbon project. For one, a carbon project (or a climate adaptation project) is an enormous waste of money if that's all we get out of it. And, for another, it's not likely sufficient motivation for the human forces of self-governance. The social sciences are a vast reservoir of knowledge about the power of human ideas to shape human history toward progressive ends. Let's put that knowledge to work. Let's create futures that are not just sustainable but inclusive of all, that end the extreme poverty and inequalities and injustices of our current worlds, that create new worlds that are worth inhabiting and in which people thrive.

Notes

1 Jasanoff recently received the 2018 Alfred O. Hirschman Prize of the Social Science Research Council, which recognises the most important theorists and engaged practitioners in the social sciences. That she follows Amartya Sen as a recipient of this award testifies to the profound significance of her work, which has done as much to redefine social science understandings of the law and democracy in technological societies as Sen's has for development and Elinor Ostrom's did for institutions for managing collective resources.
2 In its viler forms, of course, nationalism has also launched vicious anti-democratic movements, e.g. in the first half of the twentieth century, when Germany, Italy, Japan, and other nations fashioned themselves through the rationalised management of an amalgam of racialised biopolitics and state-owned industries of techno-military production. See, for example, Bauman (2000).

References

Agarwal, A. and Narain, S. (1991) Global warming in an unequal world: a case of environmental colonialism. In *Global Warming in an Unequal World: A Case of Environmental Colonialism*. Delhi, India: Centre for Science and Environment.

Allenby, B.R. and Sarewitz, D. (2011) *The Techno-Human Condition*. Cambridge, MA: MIT Press.

Anderson, B. (1991) *Imagined Communities: Reflections on the Origin and Spread of Nationalism*, 2nd edn. New York: Verso Books.

Bauman, Z. (2000) *Modernity and the Holocaust*. Ithaca, NY: Cornell University Press.

Biermann, F., Pattberg, P., Van Asselt, H., and Zelli, F. (2009) The fragmentation of global governance architectures: a framework for analysis. *Global Environmental Politics*, 9(4), 14–40.

Bulkeley, H. (2013) *Cities and Climate Change*. London, UK: Routledge.

Bulkeley, H. and Newell, P. (2015) *Governing Climate Change*. London, UK: Routledge.

Busch, L. (2011) *Standards: Recipes for Reality*. Cambridge, MA: MIT Press.

Bush, V. (1945) *Science, the Endless Frontier: A Report to the President*. Washington, DC: US GPO.

Carmen, I.H. (2004) *Politics in the Laboratory: The Constitution of Human Genomics*. Madison, WI: University of Wisconsin Press.

Chandler, A.D. (1977) *The Visible Hand: The Managerial Revolution in American Business*. Cambridge, MA: Harvard University Press.

Downey, G.L. and Zuiderent-Jerak, T. (2016) Making and doing: engagement and reflexive learning in STS. In *Handbook of Science and Technology Studies*. Cambridge, MA: MIT Press.

Eschenfelder, K.R. and Miller, C.A. (2007) Examining the role of web site information in facilitating different citizen–government relationships: a case study of state chronic wasting disease web sites. *Government Information Quarterly*, 24(1), 64–88.

Ezrahi, Y. (1990) *The Descent of Icarus: Science and the Transformation of Contemporary Democracy*. Cambridge, MA: Harvard University Press.

Ezrahi, Y. (1997) *Rubber Bullets Power and Conscience in Modern Israel*. Cambridge, UK: Cambridge University Press.

Ezrahi, Y. (2012) *Imagined Democracies: Necessary Political Fictions*. Cambridge, UK: Cambridge University Press.

Felt, U., Fouché, R., Miller, C.A., and Smith-Doerr, L. (eds) (2016) *The Handbook of Science and Technology Studies*. Cambridge, MA: MIT Press.

Foster, M.S. (2003) *A Nation on Wheels: The Automobile Culture in America since 1945*. San Francisco, CA: Thomson, Wadsworth.

Foucault, M. (1991) *The Foucault Effect: Studies in Governmentality*. Chicago, IL: University of Chicago Press.

Frischmann, B. and Selinger, E. (2018) *Re-engineering Humanity*. Cambridge, UK: Cambridge University Press.

Galambos, L. (1970) The emerging organizational synthesis in modern American history. *Business History Review*, 44(3), 279–290.

Gusterson, H. (2004) *People of the Bomb: Portraits of America's Nuclear Complex*. Minneapolis, MN: University of Minnesota Press.

Hacking, I. (1990) *The Taming of Chance*. Cambridge, UK: Cambridge University Press.

Harari, Y.N. (2016) *Homo Deus: A Brief History of Tomorrow*. New York: Random House.

Haraway, D.J. (2000) *Modest_Witness@ Second_Millennium. FemaleMan_Meets_Onco-Mouse: Feminism and Technoscience*. London, UK: Routledge.

Hays, S.P. (1959) *Conservation and the Gospel of Efficiency: The Progressive Conservation Movement, 1890–1920*. Pittsburgh, PA: University of Pittsburgh Press.

Hecht, G. (1998) *The Radiance of France: Nuclear Power and National Identity after World War II*. Cambridge, MA: MIT Press.

Herkert, J.R. (1990) Science, technology and society education for engineers. *IEEE Technology and Society Magazine*, 9(3), 22–26.

Hill, A. and Gauntlett, D. (2002) *TV Living: Television, Culture and Everyday Life*. London, UK: Routledge.

Hooghe, L., Marks, G., and Marks, G.W. (2001) *Multi-level Governance and European Integration*. New York: Rowman & Littlefield.

Hughes, T.P. (1983) *Networks of Power: Electrification in Western Society, 1880–1930*. Baltimore, MD: Johns Hopkins University Press.

Jasanoff, S. (1990) *The Fifth Branch: Science Advisers as Policymakers*. Cambridge, MA: Harvard University Press.

Jasanoff, S. (1996) *Science at the Bar*. Cambridge, MA: Harvard University Press.

Jasanoff, S. (ed.) (2004) *States of Knowledge: The Co-production of Science and the Social Order*. London, UK: Routledge.

Jasanoff, S. (2016) *The Ethics of Invention: Technology and the Human Future*. New York: WW Norton & Company.

Jasanoff, S. and Kim, S.H. (eds) (2015) *Dreamscapes of Modernity: Sociotechnical Imaginaries and the Fabrication of Power*. Chicago, IL: University of Chicago Press.

Kates, R.W., Clark, W.C., Corell, R., Hall, J.M., Jaeger, C.C., Lowe, I., McCarthy, J.J., Schellnhuber, H.J., Bolin, B., Dickson, N.,M., and Faucheux, S. (2001) Sustainability science. *Science*, 292(5517), 641–642.

Klein, N. (2015) *This Changes Everything: Capitalism vs. the Climate*. New York: Simon and Schuster.

Layton Jr, E.T. (1986) *The Revolt of the Engineers. Social Responsibility and the American Engineering Profession*. Baltimore, MD: Johns Hopkins University Press.

Leach, M., Stirling, A.C., and Scoones, I. (2010) *Dynamic Sustainabilities: Technology, Environment, Social Justice*. London, UK: Routledge.

LeMenager, S. (2014) *Living Oil: Petroleum Culture in the American Century*. Oxford, UK: Oxford University Press.

Lerner, S. (2010) *Sacrifice Zones: The Front Lines of Toxic Chemical Exposure in the United States*. Cambridge, MA: MIT Press.

Lipschutz, R.D. and Mayer, J. (1996) *Global Civil Society and Global Environmental Governance: The Politics of Nature from Place to Planet*. New York: SUNY Press.

MacKenzie, D. (2008) *An Engine, not a Camera: How Financial Models Shape Markets*. Cambridge, MA: MIT Press.

McKibben, B. (1989) *The End of Nature*. New York: Random House Incorporated.

Miller, C.A. (2015) Globalizing security: science and the transformation of contemporary political imagination. In *Dreamscapes of Modernity: Sociotechnical Imaginaries and the Fabrication of Power*. Chicago, IL: University of Chicago Press.

Miller, C.A. (2004) Climate science and the making of a global political order. In *States of Knowledge: The Coproduction of Science and Social Order*. London, UK: Routledge, 46–66.

Miller, C.A. (2010) Policy challenges and university reform. In *The Oxford Handbook of Interdisciplinarity*. Oxford, UK: Oxford University Press.

Miller, C.A., Richter, J., and O'Leary, J. (2015) Socio-energy systems design: a policy framework for energy transitions. *Energy Research & Social Science*, 6, 29–40.

Mitchell, T. (2011) *Carbon Democracy: Political Power in the Age of Oil*. New York: Verso Books.

Mumford, L. (1934) *Technics and Civilization*. Chicago, IL: University of Chicago Press.

NAE (2014) *The Climate Change Educational Partnership: Climate Change, Engineered Systems, and Society*. Washington, DC: National Academies Press.

Noble, D.F. (1979) *America by Design*. Oxford, UK: Oxford University Press.

Nowotny, H. (1990) Knowledge for certainty: Poverty, welfare institutions and the institutionalization of social science. In *Discourses on Society*. Dordrecht: Springer.

Older, M. (2017) *Infomacracy*. New York: Tor.

O'Neil, C. (2017) *Weapons of Math Destruction: How Big Data Increases Inequality and Threatens Democracy*. New York: Broadway Books.

Ostrom, E. (2010) Beyond markets and states: polycentric governance of complex economic systems. *American Economic Review*, 100(3), 641–672.

Perkins, J. (2011) *Confessions of an Economic Hit Man: The Shocking Story of How America Really Took Over the World*. New York: Random House.

Pontifical Council for Justice and Peace (2014) *Energy, Justice and Peace: A Reflection on Energy in the Current Context of Development and Environmental Protection*. The Vatican: Libreria Editrice Vaticana.

Porter, T.M. (1996) *Trust in Numbers: The Pursuit of Objectivity in Science and Public Life*. Princeton, NJ: Princeton University Press.

Rabinbach, A. (1996) Social knowledge, social risk, and the politics of industrial accidents in Germany and France. In *States, Social Knowledge, and the Origins of Modern Social Policies*. Princeton, NJ: Princeton University Press.

Rayner, S. (2010) How to eat an elephant: a bottom-up approach to climate policy. *Climate Policy*, 10(6), 615–621.

Rockström, J., Steffen, W., Noone, K., Persson, Å., Chapin III, F.S., Lambin, E., Lenton, T.M., Scheffer, M., Folke, C., Schellnhuber, H.J., and Nykvist, B. (2009) Planetary boundaries: exploring the safe operating space for humanity. *Ecology and Society*, 14(2).

Rueschemeyer, D. and Skocpol, T. (eds) (1996) *States, Social Knowledge, and the Origins of Modern Social Policies*. Princeton, NJ: Princeton University Press.

Scott, J.C. (1998) *Seeing like a State: How Certain Schemes to Improve the Human Condition have Failed*. New Haven, CT: Yale University Press.

Shelley, M., Guston, D.H., Finn, E., Robert, J.S., and Robinson, C.E. (2017) *Frankenstein: Annotated for Scientists, Engineers, and Creators of All Kinds*. Cambridge, MA: The MIT Press.

Shove, E. and Walker, G. (2014) What is energy for? Social practice and energy demand. *Theory, Culture & Society*, 31(5), 41–58.

Skocpol, T. (1992) *Protecting Soldiers and Mothers*. Cambridge, MA: Harvard University Press.

Slota, S.C. and Bowker, G.C. (2016) How infrastructures matter. In *The Handbook of Science and Technology Studies*. Cambridge, MA: MIT Press.

Smith, A. and Stirling, A. (2008) *Social-ecological Resilience and Socio-technical Transitions: Critical Issues for Sustainability Governance*. Brighton, UK: STEPS-Centre.

Soleri, P. (2012) *Lean Linear City: Arterial Arcology*. Mayer, AZ: Cosanti Press.

Taylor, P.J. and Derudder, B. (2015) *World City Network: A Global Urban Analysis*. London, UK: Routledge.

Thompson, P.B. (2010) *The Agrarian Vision: Sustainability and Environmental Ethics*. Lexington, KY: University Press of Kentucky.

Trist, E. (1981) The evolution of socio-technical systems. In *Perspectives on Organizational Design and Behavior*. New York: Wiley.

Turkle, S. (1995) *Life on the Screen*. New York: Simon and Schuster.

Vlek, C. and Steg, L. (2007) Human behavior and environmental sustainability: problems, driving forces, and research topics. *Journal of Social Issues*, 63(1), 1–19.

Voss, J.P., Bauknecht, D., and Kemp, R. (eds) (2006) *Reflexive Governance for Sustainable Development*. Cheltenham, UK: Edward Elgar Publishing.

Winner, L. (1986) *The Whale and the Reactor: A Search for Limits in an Age of High Technology*. Chicago, IL: University of Chicago Press.

Wittrock, B. and Wagner, P. (1996) Social science and the building of the early welfare state-toward a comparison of statist and non-statist societies. In *States, Social Knowledge, and the Origins of Modern Social Policies*. Princeton, NJ: Princeton University Press.

World Economic Forum (2016) *The Future of Jobs: Employment, Skills and Workforce Strategy for the Fourth Industrial Revolution*. Geneva, Switzerland: World Economic Forum.

Young, O.R. (ed.) (1997) *Global Governance: Drawing Insights from the Environmental Experience*. Cambridge, MA: MIT Press.

Young, O.R., Berkhout, F., Gallopin, G.C., Janssen, M.A., Ostrom, E. and Van der Leeuw, S. (2006) The globalization of socio-ecological systems: an agenda for scientific research. *Global Environmental Change*, 16(3), 304–316.

Index

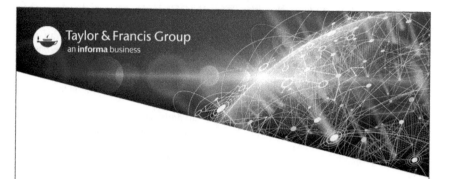